Supersymmetry and Equivariant de Rham Theory

Springer
Berlin
Heidelberg
New York
Barcelona
Hong Kong
London
Milan
Paris
Singapore
Tokyo

Victor W. Guillemin
Shlomo Sternberg

Supersymmetry and Equivariant de Rham Theory

 Springer

Victor W. Guillemin
Department of Mathematics
Massachusetts Institute
of Technology
77, Massachusetts Avenue
Cambridge, MA 02139
USA

Shlomo Sternberg
Department of Mathematics
Harvard University
One Oxford Street
Cambridge, MA 02138
USA

Jochen Brüning
Institut für Mathematik
Mathematisch-Naturwissen-
schaftliche Fakultät II
Humboldt-Universität Berlin
Unter den Linden 6
D-10117 Berlin
Germany

Cataloging-in-Publication Data applied for
Die Deutsche Bibliothek – CIP-Einheitsaufnahme
Guillemin, Victor W.:
Supersymmetry and equivariant de Rham theory / Victor W. Guillemin; Shlomo
Sternberg.- Berlin; Heidelberg; New York; Barcelona; Hong Kong; London; Milan;
Paris; Singapore; Tokyo: Springer 1999

Mathematics Subject Classification (1991): 58-XX

ISBN 978-3-642-08433-1

© Springer-Verlag Berlin Heidelberg 2010
Printed in Germany

En hommage à Henri Cartan

Preface

This is the second volume of the Springer collection *Mathematics Past and Present*. In the first volume, we republished Hörmander's fundamental papers *Fourier integral operators* together with a brief introduction written from the perspective of 1991. The composition of the second volume is somewhat different: the two papers of Cartan which are reproduced here have a total length of less than thirty pages, and the 220 page introduction which precedes them is intended not only as a commentary on these papers but as a textbook of its own, on a fascinating area of mathematics in which a lot of exciting innovations have occurred in the last few years. Thus, in this second volume the roles of the reprinted text and its commentary are reversed. The seminal ideas outlined in Cartan's two papers are taken as the point of departure for a full modern treatment of equivariant de Rham theory which does not yet exist in the literature.

We envisage that future volumes in this collection will represent both variants of the interplay between *past* and *present* mathematics: we will publish classical texts, still of vital interest, either reinterpreted against the background of fully developed theories or taken as the inspiration for original developments.

Contents

Introduction

0.1

The year 2000 will be the fiftieth anniversary of the publication of Henri Cartan's two fundamental papers on equivariant De Rham theory "Notions d'algèbre différentielle; applications aux groupes de Lie et aux variétés où opère un groupe de Lie" and "La trangression dans un groupe de Lie et dans un espace fibré principal." The aim of this monograph is to give an updated account of the material contained in these papers and to describe a few of the more exciting developments that have occurred in this area in the five decades since their appearance. This "updating" is the work of many people: of Cartan himself, of Leray, Serre, Borel, Atiyah-Bott, Berline-Vergne, Kirwan, Mathai-Quillen and others (in particular, as far as the contents of this manuscript are concerned, Hans Duistermaat, from whom we've borrowed our treatment of the Cartan isomorphism in Chapter 4, and Jaap Kalkman, whose Ph.D. thesis made us aware of the important role played by supersymmetry in this subject). As for these papers themselves, our efforts to update them have left us with a renewed admiration for the simplicity and elegance of Cartan's original exposition of this material. We predict they will be as timely in 2050 as they were fifty years ago and as they are today.

0.2

Throughout this monograph G will be a compact Lie group and g its Lie algebra. For the topologists, the *equivariant* cohomology of a G-space, M, is defined to be the ordinary cohomology of the space

$$(M \times E)/G \qquad (0.1)$$

the "E" in (0.1) being any contractible topological space on which G acts freely. We will review this definition in Chapter 1 and show that the cohomology of the space (0.1) does not depend on the choice of E.

If M is a finite-dimensional differentiable manifold there is an alternative way of defining the equivariant cohomology groups of M involving de Rham theory, and one of our goals in Chapters 2 – 4 will be to prove an equivariant version of the de Rham theorem, which asserts that these two definitions give the same answer. We will give a rough idea of how the proof of this goes:

1. Let $\xi_1, \ldots, \xi_n$ be a basis of g. If M is a differentiable manifold and the action of G on M is a differentiable action, then to each ξ_a corresponds a vector field on M and this vector field acts on the de Rham complex, $\Omega(M)$, by an "interior product" operation, ι_a, and by a "Lie differentiation" operation, L_a. These operations fit together to give a representation of the Lie superalgebra

$$\tilde{g} = g_{-1} \oplus g_0 \oplus g_1$$

g_{-1} having ι_a, $a = 1, \ldots, n$ as basis, g_0 having L_a, $a = 1, \ldots, n$ as basis and g_1 having the de Rham coboundary operator, d, as basis. The action of G on $\Omega(M)$ plus the representation of $\tilde{g}$ gives us an action on $\Omega(M)$ of the Lie *supergroup*, G^*, whose underlying manifold is G and underlying algebra is $\tilde{g}$.

Consider now the de Rham theoretic analogue of the product, $M \times E$. One would like this to be the tensor product

$$\Omega(M) \otimes \Omega(E), \tag{0.2}$$

however, it is unclear how to define $\Omega(E)$ since E has to be a contractible space on which G acts freely, and one can show such a space can not be a finite-dimensional manifold. We will show that a reasonable substitute for $\Omega(E)$ is a commutative graded superalgebra, A, equipped with a representation of G^* and having the following properties:

a. It is acyclic with respect to d. $\tag{0.3}$

b. There exist elements $\theta^b \in A^1$ satisfying $\iota_a \theta^b = \delta_a^b$.

(The first property is the de Rham theoretic substitute for the property "E is contractible" and the second for the property "G acts on E in a locally-free fashion".) Assuming such an A exists (about which we will have more to say below) we can take as our substitute for (0.2) the algebra

$$\Omega(M) \otimes A \tag{0.4}$$

As for the space (0.1), a suitable de Rham theoretic replacement is the complex

$$(\Omega(M) \otimes A)_{\text{bas}} \tag{0.5}$$

of the *basic* elements of $\Omega(M) \otimes A$, "basic" meaning G-invariant and annihilated by the ι_a's. Thus one is led to define the equivariant de Rham cohomology, of M as the cohomology of the complex (0.5). There are, of course, two things that have to be checked about this definition.

One has to check that it is independent of A, and one has to check that it gives the right answer: that the cohomology groups of the complex (0.5) are identical with the cohomology groups of the space (0.1). At the end of Chapter 2 we will show that the second statement is true provided that A is chosen appropriately: More explicitly, assume G is contained in $U(n)$ and, for $k > n$ let $\mathcal{E}_k$ be the set of orthonormal n-tuples, $(v_1, \ldots, v_n)$, with $v_i \in \mathbf{C}^k$. One has a sequence of inclusions:

$$\cdots \to \mathcal{E}_{k-1} \to \mathcal{E}_k \to \mathcal{E}_{k+1} \to \cdots$$

and a sequence of pull-back maps

$$\cdots \leftarrow \Omega(\mathcal{E}_{k-1}) \leftarrow \Omega(\mathcal{E}_k) \leftarrow \Omega(\mathcal{E}_{k+1}) \leftarrow \cdots \tag{0.6}$$

and we will show that if A is the inverse limit of this sequence, it satisfies the conditions (0.3), and with

$$E = \varprojlim \mathcal{E}_k$$

the cohomology groups of the complex (0.5) are identical with the cohomology groups of the space (0.1).

2. To show that the cohomology of the complex (0.5) is independent of A we will first show that there is a much simpler candidate for A than the "A" defined by the inverse limit of (0.6). This is the Weil algebra

$$W = \wedge(g^*) \otimes S(g^*), \tag{0.7}$$

and in Chapter 3 we will show how to equip this algebra with a representation of G^*, and show that this representation has properties (0.3), (a) and (b). Recall that the second of these two properties is the de Rham theoretic version of the property "G acts in locally free fashion on a space E". We will show that there is a nice way to formulate this property in terms of W, and this will lead us to the important notion of W^* module.

Definition 0.0.1 *A graded vector space, A, is a W^* **module** if it is both a W module and a G^* module and the map*

$$W \otimes A \longrightarrow A, \ w \otimes a \longrightarrow wa, \tag{0.8}$$

is a G^* module morphism.

3. Finally in Chapter 4 we will conclude our proof that the cohomology of the complex (0.5) is independent of A by deducing this from the following much stronger result. (See Theorem 4.3.1.)

Theorem 0.0.1 *If A is a W^* module and E an acyclic W^* algebra the G^* modules A and $A \otimes E$ have the same* basic *cohomology.*

(We will come back to another important implication of this theorem in §4 below.)

0.3

Since the cohomology of the complex (0.5) is independent of the choice of A, we can take A to be the algebra (0.7). This will give us the *Weil model* for computing the equivalent de Rham cohomology of M. In Chapter 4 we will show that this is equivalent to another model which, for computational purposes, is a lot more useful. For any G^* module, Ω, consider the tensor product

$$\Omega \otimes S(g^*) \qquad (0.9)$$

equipped with the operation

$$d(\omega \otimes f) = d\omega \otimes f - \iota_a \omega \otimes x^a f, \qquad (0.10)$$

x^a, $a = 1, \ldots, n$, being the basis of g^* dual to ξ_a, $a = 1, \ldots, n$. One can show that $d^2 = 0$ on the set of invariant elements

$$(\Omega \otimes S(g^*))^G \qquad (0.11)$$

making the space (0.11) into a cochain complex, and Cartan's theorem says that the cohomology of this complex is identical with the cohomology of the Weil model. In Chapter 4 we will give a proof of this fact based on ideas of Mathai–Quillen (with some refinements by Kalkman and ourselves). If $\Omega = \Omega(M)$ the complex, (0.10) – (0.11), is called the *Cartan model*; and many authors nowadays take the cohomology groups of this complex to be, by definition, the equivariant cohomology groups of M. From this model one can deduce (sometimes with very little effort!) lots of interesting facts about the equivariant cohomology groups of manifolds. We'll content ourselves for the moment with mentioning one: the computation of the equivariant cohomology groups of a homogeneous space. Let K be a closed subgroup of G. Then

$$H_G(G/K) \cong S(k^*)^K . \qquad (0.12)$$

(Proof: From the Cartan model it is easy to read off the identifications

$$H_G(G/K) \cong H_{G \times K}(G) \cong H_K(\text{pt.}) \qquad (0.13)$$

and it is also easy to see that the space on the far right is just $S(k^*)^K$.)

0.4

A fundamental observation of Borel [Bo] is that there exists an isomorphism

$$H_G(M) \cong H(M/G) \qquad (0.14)$$

provided G acts *freely* on M. In equivariant de Rham theory this result can easily be deduced from the theorem that we cited in Section 2 (Theorem 4.3.1

in Chapter 4). However, there is an alternative proof of this result, due to Cartan, which involves a very beautiful generalization of Chern-Weil theory: If G acts freely on M one can think of M as being a principal G-bundle with base

$$X = M/G \tag{0.15}$$

and fiber mapping

$$\pi : M \longrightarrow X \,,\; \pi(m) = m \quad \text{modulo } G\,. \tag{0.16}$$

Put a connection on this bundle and consider the map

$$\Omega(M) \otimes S(g^*) \longrightarrow \Omega(M) \tag{0.17}$$

which maps $\omega \otimes x_1^{k_1} \ldots x_n^{k_n}$ to $\omega_{\text{hor}} \otimes \mu_1^{k_1} \ldots \mu_n^{k_n}$ the μ_is being the components of the curvature form with respect to the basis, $\xi_1, \ldots, \xi_n$, of g and ω_{hor} being the horizontal component of ω. $\Omega(X)$ can be thought of as a subspace of $\Omega(M)$ via the embedding: $\Omega(X) \longrightarrow \pi^*\Omega(X)$; and one can show that the map (0.17) maps the Cartan complex (0.11) onto $\Omega(X)$. In fact one can show that this map is a cochain map and that it induces an isomorphism on cohomology. Moreover, the restriction of this map to $S(g^*)^G$ is, by definition, the Chern-Weil homomorphism. (We will prove the assertions above in Chapter 5 and will show, in fact, that they are true with $\Omega(M)$ replaced by an arbitrary W^* module.)

0.5

One important property of the Cartan complex is that it can be regarded as a *bi-complex* with bigradation

$$C^{p,q} = \left(S(g^*)^p \otimes \Omega(M)^{q-p} \right)^G \tag{0.18}$$

and the coboundary operators

$$d = 1 \otimes d_M \quad \text{and} \quad \delta = -x^a \otimes \iota_a\,. \tag{0.19}$$

This means that one can use spectral sequence techniques to compute $H_G(M)$ (or, in fact, to compute $H_G(A)$, for any G^* module, A). To avoid making "spectral sequences" a prerequisite for reading this monograph, we have included a brief review of this subject in §§ 6.1-6.4. (For simplicity we've confined ourselves to discussing the theory of spectral sequences for bicomplexes since this is the only type of spectral sequence we'll encounter.)

Applying this theory to the Cartan complex, we will show that there is a spectral sequence whose E_1 term is $H(M) \otimes S(g^*)^G$ and whose E_∞ term is $H_G(M)$. Frequently this spectral sequence collapses and when it does the (additive) equivariant cohomology of M is just

$$H(M) \otimes S(g^*)^G\,. \tag{0.20}$$

We will also use spectral sequence techniques to deduce a number of other important facts about equivariant cohomology. For instance we will show that for any G^* module, A,

$$H_G(A) \cong H_T(A)^W \tag{0.21}$$

T being the Cartan subgroup of G and W the corresponding Weyl group. We will also describe one nice topological application of (0.21): the "splitting principle" for complex vector bundles. (See [BT] page 275.)

0.6

The first half of this monograph (consisting of the sections we've just described) is basically an exegesis of Cartan's two seminal papers from 1950 on equivariant de Rham theory. In the second half we'll discuss a few of the post-1950 developments in this area. The first of these will be the Mathai-Quillen construction of a "universal" equivariant Thom form: Let V be a d-dimensional vector space and ρ a representation of G on V. We assume that ρ leaves fixed a volume form, vol, and a positive definite quadratic form $\|v\|^2$. Let $S^{\#}$ be the space of functions on V of the form, $e^{-\|v\|^2/2} p(v)$, $p(v)$ being a polynomial. In Chapter 7 we will compute the equivariant cohomology groups of the de Rham complex

$$\Omega(V)_e = S^{\#} \otimes \wedge(V^*) \tag{0.22}$$

and will show that $H_G^k(\Omega(V)_e)$ is a free $S(g^*)$-module with a single generator of degree d. We will also exhibit an example of an equivariantly closed d-form, ν, with $[\nu] \neq 0$. (This is the universal Thom form that we referred to above.) The basic ingredient in our computation is the Fermionic Fourier transform. This transform maps $\wedge(V)$ into $\wedge(V^*)$ and is defined, like the ordinary Fourier transform, by the formula

$$\int f(\psi^1, \ldots, \psi^d) \exp i\tau_k \psi^k \, d\psi \tag{0.23}$$

$\psi^1, \ldots, \psi^d$ being a basis of $\wedge^1(V)$, $\tau_1, \ldots, \tau_k$ the dual basis of $\wedge^1(V^*)$,

$$f(\psi^1, \ldots, \psi^d) = \Sigma a_I \psi^I$$

being an element of $\wedge(V)$, i.e., a "function" of the anti-commuting variables $\psi^1, \ldots, \psi^d$, and the integral being the "Berezin integral": the pairing of the integrand with the d-form vol $\in \wedge^d(V^*)$. Combining this with the usual Bosonic Fourier transform one gets a super-Fourier transform which transforms $\Omega(V)_e$ into the Koszul complex, $S(V) \otimes \wedge(V)$, and the Mathai-Quillen form into the standard generator of H_G^0 (Koszul). The inverse Fourier transform then gives one an explicit formula for the Mathai-Quillen form itself. Using the super-analogue of the fact that the restriction of the Fourier transform of a function to the origin is the integral of the function, we will get from this computation an explicit expression for the "universal" Euler class: the restriction of the universal Thom form to the origin.

0.7

Let A be a commutative G^* algebra containing $\mathbf{C}$. From the inclusion of $\mathbf{C}$ into A one gets a map on cohomology

$$H_G(\mathbf{C}) \longrightarrow H_G(A) \qquad (0.24)$$

and hence, since $H_G(\mathbf{C}) = S(g^*)^G$, a generalized Chern-Weil map:

$$S(g^*)^G \longrightarrow H_G(A). \qquad (0.25)$$

The elements in the image of this map are defined to be the "generalized characteristic classes" of A. If K is a closed subgroup of G there is a natural restriction mapping

$$H_G(A) \longrightarrow H_K(A) \qquad (0.26)$$

and under this mapping, G-characteristic classes go into K-characteristic classes. In Chapter 8 we will describe these maps in detail for the classical compact groups $U(n)$, $O(n)$ and $SO(n)$ and certain of their subgroups. Of particular importance for us will be the characteristic class associated with the element, "Pfaff", in $S(g^*)^G$ for $G = SO(2n)$. (This will play a pivotal role in the localization theorem which we'll describe below.) Specializing to vector bundles we will describe how to define the Pontryagin classes of an oriented manifold and the Chern classes of an almost complex (or symplectic) manifold, and, if M is a G-manifold, the equivariant counterparts of these classes.

0.8

Let M be a G-manifold and $\omega \in \Omega^2(M)$ a G-invariant symplectic form. A *moment map* is a G-equivariant map.

$$\phi : M \longrightarrow g^*$$

with the property that for all $\xi \in g$

$$\iota_\xi \omega = -d\phi^\xi \qquad (0.27)$$

ϕ^ξ being the ξ component of ϕ. Let ξ_i, $i = 1, \ldots, n$ be a basis of g, x^i, $i = 1, \ldots, n$ the dual basis of g^* and ϕ_i the ξ_i-th component of ϕ. The identities (0.27) can be interpreted as saying that the *equivariant* two-form

$$\tilde{\omega} = \omega - \phi_i x^i$$

is closed. This trivial fact has a number of surprisingly deep applications and we will discuss three of them in Chapter 9: the Kostant-Kirillov theorem, the Duistermaat-Heckmann theorem and its consequences, and the "minimal coupling" theorem of Sternberg. We also give a short introduction to the notion of group-valued moment map recently introduced by Alekseev, Malkin and Meinrenken [AMM].

0.9

The last two chapters of this monograph will deal with localization theorems. In Chapter 10 we will discuss the well-known Abelian localization theorem of Berline-Vergne and Atiyah-Bott and in Chapter 11 a related "abstract" localization theorem of Borel and Hsiang.

From now on we will assume that G is abelian.[1] Let M be a compact oriented d-dimensional G-manifold. The integration map

$$\int : \Omega(M) \to \mathbf{C} \tag{0.28}$$

is a morphism of G^* modules, so it induces a map on cohomology

$$\int : H_G(M) \to H_G(\mathbf{C}) = S(g^*), \tag{0.29}$$

and the localization theorem is an explicit formula for (0.29) in terms of fixed point data. If M^G is finite it asserts that

$$\int_M \mu = \sum_{p \in M^G} \frac{i_p^* \mu}{\Pi \alpha_{i,p}} \tag{0.30}$$

μ being a closed equivariant form, $i_p^* \mu$ its restriction to p and $\alpha_{i,p}$, $i = 1, \dots d$, the weights of the isotropy representation of G on the tangent space to p. (More generally, if M^G is infinite, it asserts that

$$\int_M \mu = \sum \int_{F_k} \frac{i_k^* \mu}{e_k} \tag{0.31}$$

the F_k's being the connected components of M^G and e_k being the equivariant Euler class of the normal bundle of F_k.) To prove this formula we will first of all describe how to define "push-forward" operations (or "Gysin maps") in equivariant de Rham theory; i.e., we will show that if M_1 and M_2 are G-manifolds and $f : M_1 \to M_2$ a G-map which is *proper* there is a natural "push-forward"

$$f_* : H_G^k(M_1) \to H_G^{k+\ell}(M_2), \tag{0.32}$$

ℓ being the difference between the dimension of M_2 and the dimension of M_1. To construct this map we will need to define the equivariant Thom form for a pair, (M, E), consisting of a G-manifold, M, and a vector bundle E over M on which G acts by vector bundle automorphisms; and, following Mathai-Quillen, we will show how this can be defined in terms of the *universal*

---

[1]We will prove in Chapter 10 that for a localization theorem of the form (0.31) to be true, the Euler class of the normal bundle of M^G has to be invertible, and that this more or less forces G to be Abelian. For G non-Abelian there is a more complicated localization theorem due to Witten [Wi] and Jeffrey-Kirwan [JK] in which the integration operation (0.29) gets replaced by a more subtle integration operation called "Kirwan integration".

equivariant Thom form described above. We will then show, following Atiyah-Bott, that the localization theorem is equivalent to the identity

$$i^* i_*(1) = e \qquad (0.33)$$

i being the inclusion map of M^G into M and e being the equivariant Euler class of the normal bundle of M^G.

0.10

The following theorem of Borel and Hsiang, which we will discuss in Chapter 11, is a kind of "raison d'être" for formulas of the type (0.30) and (0.31).

Theorem 0.0.2 Abstract localization theorem *The kernel of the restriction map*

$$i^* : H_G(M) \longrightarrow H_G(M^G) \qquad (0.34)$$

is the set of torsion elements in $H_G(M)$, i.e., b is in this kernel if and only if there exists a $p \in S(g^*)$ with $p \neq 0$ and $pb = 0$.

From this the identities (0.30) and (0.31) can be deduced as follows. It is clear that the map (0.29) is zero on torsion elements; so it factors through the map (0.34). In other words there is a *formal* integration operation

$$\sigma : i^* H_G(M) \longrightarrow S(g^*) \qquad (0.35)$$

whose composition with i^* is the map (0.29); and, given the fact that such an operation exists, it is not hard to deduce the formula (0.31) by checking what it does on Thom classes.

Another application of the abstract localization theorem is the following: We recall that there is a spectral sequence whose E_1 term is the tensor product (0.20) and whose E_∞ term is $H_G(M)$. Following Goresky-Kottwitz-MacPherson, we will say that M is *equivariantly formal* if this spectral sequence collapses. (See [GKM], Theorem 14.1 for a number of alternative characterizations of this property. We will discuss several of these alternative formulations in the Bibliographical Notes to Chapter 11.) If M is equivariantly formal, then by (0.20) the cohomology groups of M are

$$H_G^k(M) = \bigoplus_{i+2j=k} H^i(M) \otimes S^j(g^*), \qquad (0.36)$$

and in fact we will prove in Chapter 5 that if M is equivariantly formal,

$$H_G(M) \cong H(M) \otimes S(g^*) \qquad (0.37)$$

as an $S(g^*)$-module. We will now show that the Borel-Hsiang theorem gives one some information about the *ring* structure of $H_G(M)$. If M is equivariantly formal, then, by (0.37), $H_G(M)$ is free as an $S(g^*)$ module; so the submodule of torsion elements is $\{0\}$. Hence, by Borel-Hsiang, the map

$$i^* : H_G(M) \longrightarrow H_G(M^G)$$

is injective. However, the structure of the ring $H_G(M^G)$ is much simpler than that of $H_G(M)$; namely

$$H_G(M^G) \cong H(M^G) \otimes S(g^*); \tag{0.38}$$

so one will have more or less unraveled the ring structure of $H_G(M)$ if one can describe how the image of i^* sits inside this ring. Fortunately there is a very nice description of the image of i^*, due to Chang and Skjelbred, which says that

$$i^* H_G(M) = \bigcap_H i_H^* H_G(M^H) \tag{0.39}$$

the intersection being over all codimension-one subtori, H, of G and i_H being the inclusion map of M^G into M^H. (A proof of this using de Rham-theoretic techniques, by Michel Brion and Michele Vergne, will be given in Chapter 11.)

If one is willing to strengthen a bit the assumption of "equivalently formal" one can give a much more precise description of the right hand side of (0.39). Let us assume that M^G is finite and in addition let us make the assumption

> For every codimension-one subtorus, H, of G, $\dim M^H \leq 2$.   (0.40)

Given this assumption, one can show that there are a finite number of codimension-one subtori

$$H_i, \, i = 1, \dots, N \tag{0.41}$$

with the property

$$\dim M^{H_i} = 2,$$

and if H is not one of these exceptional subtori $M^H = M^G$. Moreover, if H is one of these exceptional subtori, the connected components, $\Sigma_{i,j}$, of M^{H_i} are 2-spheres, and each of these 2-spheres intersects M^G in exactly two points (a "north pole" and a "south pole"). For i fixed, the $\Sigma_{i,j}$'s can't intersect each other; however, for different i's, they can intersect at points of M^G; and their intersection properties can be described by an "intersection graph", Γ, whose edges are the $\Sigma_{i,j}$'s and whose vertices are the points of M^G. (Two vertices, p and q, of Γ are joined by an edge, Σ, if $\Sigma \cap M^G = \{p, q\}$.)

Moreover, for each Σ there is a unique, H_i, on the list (0.41) for which

$$\Sigma \subseteq M^{H_i}, \tag{0.42}$$

so the edges of Γ are *labeled* by the H_i's on this list.

Since M^G is finite,

$$H_G(M^G) = H^0(M^G) \otimes S(g^*) = \text{Maps}\,(M^G, S(g^*))$$

and hence

$$H_G(M^G) = \text{Maps}\,(V_\Gamma, S(g^*)) \tag{0.43}$$

where V_Γ is the set of vertices of Γ.

Theorem 0.0.3 *([GKM]) An element, p, of the ring*

$$Maps\,(V_\Gamma, S(g^*))$$

is in the image of the embedding

$$i^* : H_G(M) \longrightarrow H_G(M^G)$$

if and only if, for every edge, Σ, of the intersection graph, Γ, it satisfies the compatibility condition

$$r_h p(v_1) = r_h p(v_2) \tag{0.44}$$

v_1 and v_2 being the vertices of Σ, and h being the Lie algebra of the group (0.42) and

$$r_h : S(g^*) \longrightarrow S(h^*) \tag{0.45}$$

being the restriction map.

0.11

As we mentioned at the beginning of this introduction, the results that we've described above involve contributions by many people. The issue of provenance—who contributed what—is not easy to sort out in an area as active as this; however, we've added a bibliographical appendix to each chapter in which we attempt to set straight the historical record in so far as we can. (There is also a more personal historical record consisting of the contributions of our friends and colleagues to this project. This record is harder to set straight; however, there is one person above all to whom we would like to express our gratitude: It is to Raoul Bott that we owe our initiation into the mysteries of this subject many years ago, in the Spring of 1982 at Bures-sur Yvette just after he and Atiyah had discovered their version of the localization theorem. The ur-draft of this manuscript was twenty pages of handwritten notes based on his lectures to us at that time. We would also like to thank Matthew Leingang and Cătălin Zara for helping us to revise the first draft of this monograph and for suggesting a large number of improvements in style and content.)

Chapter 1

Equivariant Cohomology in Topology

Let G be a compact Lie group acting on a topological space X. We say that this action is **free** if, for every $p \in X$, the stabilizer group of p consists solely of the identity. In other words, the action is free if, for every $a \in G$, $a \neq e$, the action of a on X has no fixed points. If G acts freely on X then the quotient space X/G is usually as nice a topological space as X itself. For instance, if X is a manifold then so is X/G.

The definition of the equivariant cohomology group, $H_G^*(X)$ is motivated by the principle that if G acts freely on X, then the equivariant cohomology groups of X should be just the cohomology groups of X/G:

$$H_G^*(X) = H^*(X/G) \quad \text{when the action is free.} \tag{1.1}$$

For example, if we let G act on itself by left multiplication this implies that

$$H_G^*(G) = H^*(\text{pt.}). \tag{1.2}$$

If the action is not free, the space X/G might be somewhat pathological from the point of view of cohomology theory. Then the idea is that $H_G^*(X)$ is the "correct" substitute for $H^*(X/G)$.

1.1 Equivariant Cohomology via Classifying Bundles

Cohomology is unchanged by homotopy equivalence. So our motivating principle suggests that the equivariant cohomology of X should be the ordinary cohomology of X^*/G where X^* is a topological space homotopy equivalent to X and on which G does act freely. The standard way of constructing such a space is to take it to be the product $X^* = X \times E$ where E is a contractible space on which G acts freely. Thus the standard way of defining the equivariant cohomology groups of X is by the recipe

$$H_G^*(X) := H^* \left((X \times E)/G \right). \tag{1.3}$$

We will discuss the legitimacy of this definition below. We must show that it does not depend on the choice of E. Before doing so we note that if G acts freely on X then the projection

$$X \times E \to X$$

onto the first factor gives rise to a map

$$(X \times E)/G \to X/G \tag{1.4}$$

which is a fibration with typical fiber E. Since E is contractible we conclude that

$$H_G^*(X) = H^* \left((X \times E)/G \right) = H^*(X/G),$$

in compliance with (1.1). Notice also that since (1.4) is a fiber bundle over X/G with contractible fiber, it admits a global cross-section

$$s : X/G \to (X \times E)/G. \tag{1.5}$$

The projection

$$X \times E \to E$$

onto the second factor gives rise to a map

$$(X \times E)/G \to E/G. \tag{1.6}$$

Composing (1.6) with the section s gives rise to a map

$$f : X/G \to E/G. \tag{1.7}$$

Let

$$\begin{aligned} \pi : X &\to X/G \\ \rho : E &\to E/G \end{aligned}$$

be the projections of X and E onto their quotient spaces under the respective G-actions.

Proposition 1.1.1 *Suppose that G acts freely on X and that E is a contractible space on which G acts freely. Any cross-section $s : X/G \to (X \times E)/G$ determines a unique G-equivariant map*

$$h : X \to E$$

which makes the diagram

$$X \xrightarrow{\quad h \quad} E$$

$$\pi \downarrow \qquad \qquad \downarrow \rho$$

$$X/G \xrightarrow{\quad f \quad} E/G \qquad\qquad (1.8)$$

commute. Conversely, every G-equivariant map $h : X \to E$ determines a section $s : X/G \to (X \times E)/G$ and a map f which makes (1.8) commute. Any two such sections are homotopic and hence the homotopy class of (f,h) is unique, independent of the choice of s.

Proof. Let $y \in X/G$ and consider the preimage of y in $(X \times E)/G$. This preimage consists of all pairs

$$(x,e) \in X \times E, \quad \pi(x) = y$$

modulo the equivalence relation

$$(x,e) \sim (ax,ae), \quad a \in G.$$

Each such equivalence class can be thought of as the graph of a G-equivariant map

$$\pi^{-1}(y) \to E.$$

So $s(y)$ determines such a map for every y. In other words we have defined $h : X \to E$ by the formula

$$s(y) = [(x,h(x))]$$

where $[\ ]$ denotes equivalence class modulo G. The definition (1.7) of f then says that the square (1.8) commutes. Since the fibers of $(X \times E)/G \to X/G$ are contractible, any two cross-sections are homotopic, proving the last assertion in the proposition. $\square$

Proposition 1.1.1 is usually stated as a theorem about principal bundles: Since G acts freely on X we can consider X as a principal bundle over

$$Y := X/G.$$

Similarly we can regard E as a principal bundle over

$$B := E/G.$$

Proposition 1.1.1 is then equivalent to the following "classification theorem" for principal bundles:

Theorem 1.1.1 *Let Y be a topological space and $\pi : X \to Y$ a principal G-bundle. Then there exists a map*

$$f : Y \to B$$

and an isomorphism of principal bundles

$$\Phi : X \to f^*E$$

where f^*E is the "pull-back" of the bundle $E \to B$ to X. Moreover f and Φ are unique up to homotopy.

Remarks.

1. $f^*E = \{(y,e)|\ f(y) = \rho(e)\}$ so the projection $(y,e) \mapsto y$ makes f^*E into a principal G-bundle over X. This is the construction of the pull-back bundle.

2. We can reformulate Theorem 1.1.1 as saying that there is a one-to-one correspondence between equivalence classes of principal G-bundles and homotopy classes of mappings $f : Y \to B$. In other words, Theorem 1.1.1 reduces the classification problem for principal G bundles over Y to the homotopy problem of classifying maps of Y into B up to homotopy. For this reason the space B is called the *classifying space* for G and the bundle $E \to B$ is called the *classifying bundle* .

One important consequence of Theorem 1.1.1 is:

Theorem 1.1.2 *If E_1 and E_2 are contractible spaces on which G acts freely, they are equivalent as G-spaces. In other words there exist G-equivariant maps*

$$\phi : E_1 \to E_2, \quad \psi : E_2 \to E_1$$

with G-equivariant homotopies

$$\psi \circ \phi \simeq id_{E_1}, \quad \phi \circ \psi \simeq id \ _{E_2}.$$

Proof.   The existence of ϕ follows from Theorem 1.1.1 with $X = E_1$ and $E = E_2$. Similarly the existence of ψ follows from Theorem 1.1.1 with $X = E_2$ and $E = E_1$. Both id_{E_1} and $\psi \circ \phi$ are maps of $E_1 \to E_1$ satisfying the conditions of Theorem 1.1.1 and so are homotopic to one another. Similarly for the homotopy $\phi \circ \psi \sim id_{E_2}$. $\Box$
    A consequence of Theorem 1.1.2 is:

Theorem 1.1.3 *The definition (1.3) is independent of the choice of E.*

1.2   Existence of Classifying Spaces

Theorem 1.1.3 says that our definition of equivariant cohomology does not depend on which E we choose. But does such an E exist? In other words, given a compact Lie group G can we find a contractible space E on which G acts freely? If G is a subgroup of the compact Lie group K and we have found an E that "works" for K, then restricting the K-action to the subgroup G produces a free G-action. Every compact Lie group has a faithful linear representation, which means that it can be embedded as a subgroup of $U(n)$ for large enough n. So it is enough for us to construct a space E which is contractible and on which $U(n)$ acts freely.

Let V be an infinite dimensional separable Hilbert space. To be precise, take

$$V = L^2[0, \infty),$$

the space of square integrable functions on the positive real numbers relative to Lebesgue measure. But of course all separable Hilbert spaces are isomorphic.

Let E consist of the set of all n–tuples

$$\mathbf{v} = (v_1, \ldots, v_n) \quad v_i \in V, \quad (v_i, v_j) = \delta_{ij}.$$

The group $U(n)$ acts on E by

$$A\mathbf{v} = \mathbf{w} = (w_1, \ldots, w_n), \quad w_i = \sum_j a_{ij} v_j. \tag{1.9}$$

This action is clearly free.

So we will have proved the existence of classifying spaces for any compact Lie group once we prove:

Proposition 1.2.1 *The space E is contractible.*

We reduce the proof to two steps. To emphasize that we are working within the model where $V = L^2[0, \infty)$ we will denote elements of V by f or g. Let $E' \subset E$ consist of n-tuples of functions which all vanish on the interval $[0, 1]$.

Lemma 1.2.1 *There is a deformation retract of E onto E'.*

Proof. For any $f \in V$ define $T_t f$ by

$$T_t f(x) = \begin{cases} 0 & \text{for} \quad 0 \le x < t; \\ f(x - t) & \text{for} \quad t \le x < \infty. \end{cases}$$

Define

$$\mathbf{T}_t \mathbf{f} = (T_t f_1, \ldots, T_t f_n), \quad \text{for } \mathbf{f} = (f_1, \ldots, f_n).$$

Since T_t preserves scalar products we see that $\mathbf{T}_t$ is a deformation retract of E onto E'. $\square$

Notice that every component of $\mathbf{f}$ is orthogonal in V to any function $g \in V$ which is supported in $[0, 1]$. Therefore if $\mathbf{f} \in E'$ and $\mathbf{g} \in E$ has all its components supported in $[0, 1]$ the "rotated frame" given by

$$\mathbf{r}_t \mathbf{f} := \left((\cos \frac{\pi}{2} t) f_1 + (\sin \frac{\pi}{2} t) g_1, \ldots, (\cos \frac{\pi}{2} t) f_n + (\sin \frac{\pi}{2} t) g_n \right)$$

belongs to E for all t.

Lemma 1.2.2 E' *is contractible to a point within* E.

Proof. Pick a point $\mathbf{g}$ all of whose components are supported in $[0, 1]$. Then for any $\mathbf{f} \in E'$ the curve $\mathbf{r}_t \mathbf{f}$ as defined above starts at $\mathbf{f}$ when $t = 0$ and ends at $\mathbf{g}$ when $t = 1$. $\square$

1.3   Bibliographical Notes for Chapter 1

1. The definition (1.3) and most of the results outlined in this chapter are due to Borel (See [Bo]). The proof we've given of the contractibility of the space of orthonormal n-frames in $L^2[0, \infty)$ is related to Kuiper's proof ([Ku]) of the contractibility of the unitary group of Hilbert space.

2. The space E that we have constructed is not finite-dimensional, in particular not a finite-dimensional manifold. In order to obtain an object which can play the role of E in de Rham theory, we will be forced to reformulate some of the properties of G-actions on manifolds, like "free-ness" and "contractibility" in a more algebraic language. Having done this (in chapter 2), we will come back to the question of how to give a de Rham theoretic definition of the cohomology groups $H_G^*(M)$.

3. Let $\mathcal{C}$ be a category of topological spaces (e.g. differentiable manifolds, finite CW complexes, ...). A topological space E is said to be **contractible with respect to** $\mathcal{C}$ if, for $X \in \mathcal{C}$, every continuous map of X into E is contractible to a point. In our definition (1.3) one can weaken the assumption that E be contractible. If $X \in \mathcal{C}$ it suffices to assume that E is contractible with respect to $\mathcal{C}$. (It's easy to see that the proof of the theorems of this chapter are unaffected by this assumption.)

4. For the category $\mathcal{C}$ of finite dimensional manifolds a standard choice of E is the direct limit

$$\lim_{k \to \infty} \mathcal{E}_k,$$

$\mathcal{E}_k$ being the space of orthonormal n-frames in $\mathbf{C}^{k+1}$, $k \geq n$. This space has a slightly nicer topology than does the "E" described in section 1.2. Moreover, even though this space is not a finite-dimensional manifold, it does have a nice de Rham complex. In fact, for any finite-dimensional manifold, X, we will be able to define the de Rham complex of $(X \times E)/G$ and hence give a de Rham-theoretic definition of the cohomology groups (1.3). The details will be described in Chapter 2.

5. For $G = S^1$, $\mathcal{E}_k$ is just the $(2k+1)$-sphere

$$S^{2k+1} = \left\{ z \in \mathbf{C}^{k+1}, \, |z_0|^2 + \cdots + |z_k|^2 = 1 \right\}.$$

Consider the map of S^{2k+1} onto the standard k-simplex

$$\gamma : S^{2k+1} \to \Delta_k \qquad z \mapsto (|z_0|^2, \cdots, |z_k|^2).$$

One can reconstruct S^{2k+1} from this map by considering the relation: $z \sim z'$ iff $\gamma(z) = \gamma(z')$. This gives one a description of S^{2k+1} as the product

$$(S^1)^{k+1} \times \Delta_k = \left\{ (z_0, t_0), \cdots, (z_k, t_k) \, ; \, z_i \in S^1 \, , \, t \in \Delta_k \right\}$$

modulo the identifications

$$(z, t) \sim (z', t') \text{ iff } t_i = t'_i \text{ and } z_i = z'_i \text{ where } t_i \neq 0 \, .$$

Milnor observed that if one replaces S^1 by G in this construction, one gets a topological space $\mathcal{E}_G^k$ on which G acts freely (by its diagonal action on G^{k+1}). Moreover, he proves that if X is a finite CW-complex, then, for k sufficiently large, every continuous map of X into $\mathcal{E}_G^k$ is contractible to a point. (For more about this beautiful construction see [Mi].)

6. Except for the material that we have already covered in this chapter, the rest of the book will be devoted to the study of the equivariant cohomology groups of manifolds as defined by Cartan and Weil using equivariant de Rham theory. In particular, we will be essentially ignoring the purely *topological* side of the subject, in which the objects studied are arbitrary topological spaces X with group actions, and $H_G(X)$ is defined by the method of Borel as described in this chapter. For an introduction to the topological side of the subject, the two basic classical references are [Bo] and [Hs]. A very good modern treatment of the subject is to be found in [AP].

Chapter 2

$G^\star$ Modules

Throughout the rest of this monograph we will use a restricted version of the Einstein summation convention : A summation is implied whenever a repeated *Latin* letter occurs as a superscript and a subscript, but not if the repeated index is a Greek letter. So, for example, if g is a Lie algebra, and we have fixed a basis, $\xi_1, \ldots, \xi_n$ of g, we have

$$[\xi_i, \xi_j] = c_{ij}^k \xi_k$$

where the c_{ij}^k are called the **structure constants** of g relative to our chosen basis.

2.1 Differential-Geometric Identities

Let G be a Lie group with Lie algebra g, and suppose that we are given a smooth action of G on a differentiable manifold M. So to each $a \in G$ we have a smooth transformation

$$\phi_a : M \to M$$

such that

$$\phi_{ab} = \phi_a \circ \phi_b.$$

Let $\Omega(M)$ denote the de Rham complex of M, i.e., the ring of differential forms together with the operator d. We get a representation $\rho = \rho^M$ of G on $\Omega(M)$ where

$$\rho_a \omega = (\phi_a^{-1})^* \omega, \quad a \in G, \quad \omega \in \Omega(M).$$

We will usually drop the symbol ρ and simply write

$$a\omega = (\phi_a^{-1})^* \omega.$$

We get a corresponding representation of the Lie algebra g of G which we denote by $\xi \mapsto L_\xi$, where

$$L_\xi \omega := \frac{d}{dt} \left(\rho_{\exp t\xi}\omega\right)\bigg|_{t=0} = \frac{d}{dt} \left((\exp -t\xi)^*\omega\right)\bigg|_{t=0}, \quad \xi \in g, \ \omega \in \Omega(M). \quad (2.1)$$

The operator $L_\xi : \Omega(M) \to \Omega(M)$ is an *even* derivation (more precisely, a derivation of degree zero) in that

$$L_\xi : \Omega^k(M) \to \Omega^k(M)$$

and

$$L_\xi(\mu\nu) = (L_\xi\mu)\nu + \mu(L_\xi\nu)$$

where we have dropped the usual wedge product sign in the multiplication in $\Omega(M)$.

Let us be explicit about the convention we are using in (2.1) and will follow hereafter: The element $\xi \in g$ defines a one parameter subgroup $t \mapsto \exp t\xi$ of G, and hence the action of G on M restricts to an action of this one parameter on M. This one parameter group of transformations has an "infinitesimal generator", that is, a vector field which generates it. We may denote this vector field by ξ_M^+ so that the value of ξ_M^+ at $x \in M$ is given by

$$\xi_M^+(x) := \frac{d}{dt} \left(\exp t\xi\right)(x)\bigg|_{t=0}.$$

However the representation ρ_a is given by $\rho_a\omega = \phi_a^{-1*}\omega$ and hence, to get an action of g on $\Omega(M)$ we must consider the Lie derivative with respect to the infinitesimal generator of the one parameter group $t \mapsto \exp(-t\xi)$, which is the vector field

$$\xi_M^- = -\xi_M^+.$$

We will call *this* vector field the "vector field corresponding to ξ on M," and, as above, write L_ξ for the Lie derivative with respect to this vector field, instead of the more awkward $L_{\xi_M^-}$.

We also have the operation of interior product by the vector field corresponding to ξ. We denote it by ι_ξ. So, for each $\xi \in g$,

$$\iota_\xi : \Omega^k(M) \to \Omega^{k-1}(M)$$

and is an *odd* derivation (more precisely, a derivation of degree -1) in the sense that

$$\iota_\xi(\mu\nu) = (\iota_\xi\mu)\nu + (-1)^m\mu(\iota_\xi\nu), \quad \text{if } \mu \in \Omega^m(M).$$

Finally, we have the exterior differential

$$d : \Omega^k(M) \to \Omega^{k+1}(M)$$

which is an *odd* derivation (of degree +1) in that

$$d(\mu\nu) = (d\mu)\nu + (-1)^m \mu(d\nu) \quad \text{if } \mu \in \Omega^m(M).$$

These operators satisfy the following fundamental differential-geometric identities (the Weil equations):

$$\iota_\xi \iota_\eta + \iota_\eta \iota_\xi = 0, \tag{2.2}$$
$$L_\xi \iota_\eta - \iota_\eta L_\xi = \iota_{[\xi,\eta]}, \tag{2.3}$$
$$L_\xi L_\eta - L_\eta L_\xi = L_{[\xi,\eta]}, \tag{2.4}$$
$$d\iota_\xi + \iota_\xi d = L_\xi, \tag{2.5}$$
$$dL_\xi - L_\xi d = 0, \tag{2.6}$$
$$d^2 = 0. \tag{2.7}$$

Furthermore,

$$\rho_a \circ L_\xi \circ \rho_a^{-1} = L_{\mathrm{Ad}_a \xi} \tag{2.8}$$

and

$$\rho_a \circ \iota_\xi \circ \rho_a^{-1} = \iota_{\mathrm{Ad}_a \xi} \tag{2.9}$$

where Ad denotes the adjoint representation of G on g. In terms of our basis, we will always use the shortcut notation

$$L_j := L_{\xi_j} \quad \text{and} \quad \iota_j := \iota_{\xi_j}$$

and so can write equations (2.2)-(2.7) as

$$\iota_i \iota_j + \iota_j \iota_i = 0, \tag{2.10}$$
$$L_i \iota_j - \iota_j L_i = c_{ij}^k \iota_k, \tag{2.11}$$
$$L_i L_j - L_j L_i = c_{ij}^k L_k, \tag{2.12}$$
$$d\iota_i + \iota_i d = L_i, \tag{2.13}$$
$$dL_i - L_i d = 0 \tag{2.14}$$
$$d^2 = 0. \tag{2.15}$$

One of the key ideas of Cartan's papers was to regard these identities as being more or less the definition of a G-action on M. Nowadays, we would use the language of "super" mathematics and express equations (2.2)–(2.7) or (2.10)–(2.15) as defining a Lie superalgebra. We pause to review this language.

2.2 The Language of Superalgebra

In the world of "super" mathematics all vector spaces and algebras are graded over $\mathbf{Z}/2\mathbf{Z}$. So a **supervector space**, or simply a vector space is a vector space V with a $\mathbf{Z}/2\mathbf{Z}$ gradation:

$$V = V_0 \oplus V_1$$

where $\mathbf{Z}/2\mathbf{Z} = \{0,1\}$ in the obvious notation. An element of V_0 is called **even**, and an element of V_1 is called **odd**. Most of the time, our vector spaces will come equipped with a $\mathbf{Z}$-gradation

$$V = \bigoplus_{i \in \mathbf{Z}} V_i$$

in which case it is understood that an element of V_{2j} is even:

$$V_0 := \bigoplus V_{2j}$$

and an element of V_{2j+1} is odd:

$$V_1 := \bigoplus V_{2j+1}.$$

An element of V_i is said to have **degree** i.

A **superalgebra** (or just algebra) is a supervector space A with a multiplication satisfying

$$A_\mathbf{i} \cdot A_\mathbf{j} \subset A_{\mathbf{i+j}},$$

or

$$A_i \cdot A_j \subset A_{i+j}$$

if A is $\mathbf{Z}$-graded. For example, if V is a supervector space then $\operatorname{End} V$ is a superalgebra where

$$(\operatorname{End} V)_\mathbf{i} := \{A \in \operatorname{End} V \,|\, A : V_\mathbf{j} \to V_{\mathbf{j+i}}\}$$

or

$$(\operatorname{End} V)_i := \{A \in \operatorname{End} V \,|\, A : V_j \to V_{j+i}\}$$

in the $\mathbf{Z}$-graded case, if only finitely many of the $V_i \neq \{0\}$ (which will frequently be the case in our applications). We will also write $\operatorname{End}_i(V)$ instead of $(\operatorname{End} V)_i$ as a more pleasant notation. (In the case that infinitely many of the $V_i \neq 0$, $\operatorname{End} V$ is not the direct sum of the $\operatorname{End}_i V$: an element of $\operatorname{End} V$ might, for example, have infinitely many different degrees even if it were homogeneous on each V_i. In this case, we define

$$\operatorname{End}_{\mathbf{Z}} V := \bigoplus \operatorname{End}_i V.)$$

The basic rule in supermathematics (Quillen's law) is that all definitions which involve moving one symbol past another (in ordinary mathematics) cost a sign when both symbols are odd in supermathematics. We now turn to a list of examples of Quillen's law, all of which we will use later on:

Examples.

- The supercommutator (or just the *commutator*) of two endomorphisms of a (super)vector space is defined as

$$[L, M] := LM - (-1)^{ij} ML \quad \text{if } L \in \text{End}_i V, \quad M \in \text{End}_j V. \quad (2.16)$$

We now recognize all the expressions on the left hand sides of equations (2.2)–(2.7) as commutators. More generally, we define the commutator of any two elements in any associative superalgebra A in exactly the same way:

$$[L, M] := LM - (-1)^{ij} ML \quad \text{if } L \in A_i, \quad M \in A_j.$$

- An associative algebra is called **(super)commutative** if the commutator of any two elements vanishes. So, for example, the algebra $\Omega(M)$ of all differential forms on a manifold is a commutative superalgebra.

- A **(Z-graded) Lie superalgebra** is a **Z**-graded vector space

$$h = \bigoplus_{i \in \mathbf{Z}} h_i$$

equipped with a bracket operation

$$[,] : h_i \times h_j \to h_{i+j}$$

which is (super) anticommutative in the sense that

$$[u, v] + (-1)^{ij} [v, u] = 0, \quad \forall \, u \in h_i, \; v \in h_j,$$

and satisfies the **super version of the Jacobi identity**

$$[u, [v, w]] = [[u, v], w] + (-1)^{ij} [v, [u, w]], \quad \forall \, u \in h_i, \; v \in h_j.$$

For example, if g is an ordinary Lie algebra in the old-fashioned sense, and we have chosen a basis, $\xi_1, \ldots, \xi_n$ of g, define $\tilde{g}$ to be the Lie superalgebra

$$\tilde{g} := g_{-1} \oplus g_0 \oplus g_1$$

where g_{-1} is an n-dimensional vector space with basis $\iota_1, \ldots, \iota_n$, where g_0 is an n-dimensional vector space with basis $L_1, \ldots, L_n$ and where g_1 is a one-dimensional vector space with basis d. The bracket is defined in terms of this basis by

$$
\begin{aligned}
[\iota_a, \iota_b] &= 0, & (2.17) \\
[L_a, \iota_b] &= c_{ab}^k \iota_k & (2.18) \\
[L_a, L_b] &= c_{ab}^k L_k & (2.19) \\
[d, \iota_a] &= L_a, & (2.20) \\
[d, L_a] &= 0 & (2.21) \\
[d, d] &= 0. & (2.22)
\end{aligned}
$$

Notice that this is just a transcription of (2.10)–(2.15) with commutators replaced by brackets.

The Lie superalgebra, $\tilde{g}$, will be the fundamental object in the rest of this monograph. We repeat its definition in basis-free language: The assertion

$$\tilde{g} = g_{-1} \oplus g_0 \oplus g_1$$

as a **Z**-graded algebra implies that

$$[g_{-1}, g_{-1}] = 0 \quad \text{and} \quad [g_1, g_1] = 0.$$

The subalgebra g_0 is isomorphic to g; we if we denote the typical element of g_0 by L_ξ, $\xi \in g$, then

$$[L_\xi, L_\eta] = L_{[\xi,\eta]}$$

gives the bracket $[\,,\,] : g_0 \times g_0 \to g_0$. The space g_{-1} is isomorphic to g as a vector space, and $[\,] : g_0 \times g_{-1} \to g_{-1}$ is the adjoint representation: if we denote an element of g_{-1} by ι_η, $\eta \in g$ then

$$[L_\xi, \iota_\eta] = \iota_{[\xi,\eta]}.$$

The bracket $[\,,\,] : g_0 \times g_1 \to g_1$ is 0, and the bracket $[\,,\,] : g_{-1} \times g_1 \to g_0$ is given by

$$[\iota_\xi, d] = L_\xi.$$

- If A is a superalgebra (not necessarily associative) then Der A is the subspace of End A where

$$\mathrm{Der}_k A \subset \mathrm{End}_k A$$

consists of those endomorphisms D which satisfy

$$D(uv) = (Du)v + (-1)^{\mathbf{km}} u(Dv), \quad \text{when } u \in A_{\mathbf{m}}.$$

Similarly for the **Z**-graded case. An element of $\mathrm{Der}_k A$ is called a *derivation* of degree **k**, even or odd as the case may be.

For example, in the geometric situation studied in the preceding section, the elements of g_i act as derivations of degree i on $\Omega(M)$. So we can formulate equations (2.10)–(2.15) as saying that the Lie superalgebra, $\tilde{g}$ acts as derivations on the commutative algebra $A = \Omega(M)$ whenever we are given an action of G on M.

A second important example of a derivation is bracket by an element in a Lie superalgebra. Indeed, the super version of the Jacobi identity given above can be formulated as saying that for any fixed $u \in h_i$, the map

$$v \mapsto [u, v]$$

of the Lie superalgebra h into itself is a derivation of degree i.

- Four important facts about derivations are used repeatedly:

1) if two derivations agree on a system of generators of an algebra, they agree throughout; and

2) The field of scalars lies in A_0 and $D\alpha = 0$ if D is a derivation and α is a scalar since $D1 = D1^2 = 2D1$ and our field is not of characteristic two.

3) Der A is a Lie subalgebra of End A under commutator brackets, i.e. the commutator of two derivations is again a derivation. We illustrate by proving this last assertion for the case of two odd derivations, d_1 and d_2: Let u be an element of degree m. We have

$$
\begin{aligned}
d_1 d_2(uv) &= d_1[(d_2 u)v + (-1)^m u d_2 v] \\
&= (d_1 d_2 u)v + (-1)^{m+1} d_2 u d_1 v + (-1)^m d_1 u d_2 v + u d_1 d_2 v.
\end{aligned}
$$

Interchanging d_1 and d_2 and adding gives

$$
\begin{aligned}
[d_1, d_2](uv) &= (d_1 d_2 + d_2 d_1)(uv) \\
&= ([d_1, d_2]u)v + u[d_1, d_2]v.
\end{aligned}
$$

In particular, the square of an odd derivation is an even derivation. So, by combining 1) and 3):

4) If D is an odd derivation, to verify that $D^2 = 0$, it is enough to check this on generators.

- An ungraded algebra can be considered as a superalgebra by declaring that all its (non-zero) elements are even and there are no non-zero elements of odd degree.

- An ordinary algebra which is graded over $\mathbf{Z}$ can be made into a superalgebra with only even non-zero elements by doubling the original degrees of every element. If the original algebra was commutative in the ordinary sense, this superalgebra (with only even non-zero elements) is supercommutative. An example that we will use frequently is the symmetric algebra, $S(V)$ of an ordinary vector space, V. We may think of an element of $S^k(V)$ as a homogeneous polynomial of ordinary degree k on V^*. But we assign degree $2k$ to such an element in our supermathematical setting. Then $S(V)$ becomes a commutative superalgebra. Similarly, an ordinary Lie algebra which is graded over $\mathbf{Z}$ becomes a Lie superalgebra by doubling the degree of every element.

- If A and B are (super) algebras, the product law on $A \otimes B$ is defined by

$$
(a_1 \otimes b_1) \cdot (a_2 \otimes b_2) = (-1)^{ij} a_1 a_2 \otimes b_1 b_2
$$

where deg $a_2 = i$ and deg $b_1 = j$. With this definition, the tensor product of two commutative algebras is again commutative. Our definition of multiplication is the unique definition such that the maps

$$A \to A \otimes B \qquad a \mapsto a \otimes 1$$
$$B \to A \otimes B \qquad b \mapsto 1 \otimes b$$

are algebra monomorphisms and such that

$$(a \otimes 1) \cdot (1 \otimes b) = a \otimes b.$$

For example, let V and W be (ordinary) vector spaces. We can choose a basis $e_1 \ldots, e_m, f_1 \ldots f_n$ of $V \oplus W$ with the $e_i \in V$ and the $f_j \in W$. Thus monomials of the form

$$e_{i_1} \wedge \cdots \wedge e_{i_k} \wedge f_{j_1} \wedge \cdots f_{j_\ell}$$

constitute a basis of $\wedge(V \oplus W)$. This shows that in our category of superalgebras we have $\wedge(V \oplus W) = \wedge(V) \otimes \wedge(W)$. If M and N are smooth manifolds, then $\Omega(M) \otimes \Omega(N)$ is a subalgebra of $\Omega(M \times N)$ which is dense in the C^∞ topology.

- Our definition of the tensor product of two superalgebras and the attendant multiplication has the following universal property: Let

$$u : A \to C, \qquad v : B \to C$$

be morphisms of superalgebras such that

$$[u(a), v(b)] = 0, \qquad \forall a \in A, b \in B.$$

Then there exists a unique superalgebra morphism

$$w : A \otimes B \to C$$

such that

$$w(a \otimes 1) = u(a), \qquad w(1 \otimes b) = v(b).$$

- If V and W are supervector spaces, we can regard $\mathrm{End}(V) \otimes \mathrm{End}(W)$ as a subspace of $\mathrm{End}(V \otimes W)$ according to the rule

$$(a \otimes b)(x \otimes y) = (-1)^{qp} ax \otimes by, \qquad \deg b = q, \qquad \deg x = p.$$

Our law for the tensor product of two algebras ensures that

$$\mathrm{End}\, V \otimes \mathrm{End}\, W$$

is, in fact, a subalgebra of $\mathrm{End}(V \otimes W)$. Indeed,

$$
\begin{aligned}
(a_1 \otimes b_1)\left((a_2 \otimes b_2)(x \otimes y)\right) &= (-1)^{pq}(a_1 \otimes b_1)(a_2 x \otimes b_2 y) \\
&= (-1)^{pq}(-1)^{j(p+i)} a_1 a_2 x \otimes b_1 b_2 y
\end{aligned}
$$

where $\deg x = p$, $\deg b_2 = q$, $\deg b_1 = j$ and $\deg a_2 = i$, while

$$
\begin{aligned}
\left((a_1 \otimes b_1)(a_2 \otimes b_2)\right)(x \otimes y) &= (-1)^{ij}(a_1 a_2 \otimes b_1 b_2)(x \otimes y) \\
&= (-1)^{ij}(-1)^{(j+q)p} a_1 a_2 x \otimes b_1 b_2 y
\end{aligned}
$$

so the multiplication on $\mathrm{End}(V \otimes W)$ restricts to that of $\mathrm{End}\, V \otimes \mathrm{End}\, W$.

2.3   From Geometry to Algebra

Motivated by the geometric example, where G is a Lie group acting on a manifold, and $A = \Omega(M)$ with the Lie derivatives and interior products as described above, we make the following general definition: Let G be any Lie group, let g be its Lie algebra, and $\tilde{g}$ the corresponding Lie superalgebra as constructed above.

Definition 2.3.1 *A G^* algebra is a commutative superalgebra A, together with a representation ρ of G as automorphisms of A and an action of $\tilde{g}$ as (super)derivations of A which are consistent in the sense that*

$$\left. \frac{d}{dt}\rho(\exp t\xi)\right|_{t=0} = L_\xi \qquad (2.23)$$

$$\rho(a)L_\xi\rho(a^{-1}) = L_{\mathrm{Ad}_a \xi} \qquad (2.24)$$

$$\rho(a)\iota_\xi\rho(a^{-1}) = \iota_{\mathrm{Ad}_a \xi} \qquad (2.25)$$

$$\rho(a)d\rho(a^{-1}) = d \qquad (2.26)$$

for all $a \in G$, $\xi \in g$.

A G^* **module** *is a supervector space A together with a linear representation of G on A and a homomorphism $\tilde{g} \to \mathrm{End}(A)$ such that (2.23)–(2.26) hold. So a G^* algebra is a commutative superalgebra which is a G^* module with the additional condition that G acts as algebra automorphisms and $\tilde{g}$ acts as superderivations.*

Remarks.

1. In order for (2.23) to make sense the derivative occurring on the left side of (2.23) has to be defined. This we can do either by assuming that A possesses some kind of topology or by assuming that every element of A is G-finite, i.e. is contained in a finite G-invariant subspace of A. An example of an algebra of the first type is the de Rham complex $\Omega(M)$, and of the second type is the symmetric algebra $S(g^*) = \oplus S^i(g^*)$. (The tensor product $\Omega(M) \otimes S(g^*)$, which will figure prominently in our discussion of the Cartan model in chapter 4, is an amalgam of an algebra of the first type and the second type.)

2. This question of A having a topology (or being generated by its G-finite elements) will also come up in the next section when we consider the averaging operator

$$a \in A \longmapsto \int_G \rho(g)a\, dg$$

dg being the Haar measure.

3. If A doesn't have a topology one should, strictly speaking, qualify every assertion involving the differentiation operation (2.23) or the integration operation by adding the phrase "for G-finite elements of A"; however, we will deliberately be a bit sloppy about this.

4. Notice that if G is connected, the last three conditions, (2.24)–(2.26), are consequences of the first condition, (2.23). For example, to verify (2.25) in the connected case, it is enough to verify it for a of the form $a = \exp t\zeta$, $\zeta \in g$. It follows from (2.23) that

$$\frac{d}{dt}\rho(\exp t\zeta) = \rho(\exp t\zeta) \circ L_\zeta$$

for all t, and hence

$$\frac{d}{dt}\left[\rho(\exp t\zeta) \circ \iota_{\text{Ad}(\exp -t\zeta)\eta} \circ \rho(\exp -t\zeta)\right] =$$

$$\rho(\exp t\zeta) \circ \left([L_\zeta, \iota_{\text{Ad}(\exp -t\zeta)\eta}] - \iota_{[\zeta,\text{Ad}(\exp -t\zeta)\eta]}\right) \circ \rho(\exp -t\zeta) = 0$$

by the fact that we have an action of $\tilde{g}$. Taking $a = \exp t\zeta$ and $\xi = \text{Ad}_{a^{-1}} \eta$ proves (2.25). A similar argument proves (2.26).

5. Clearly a G^* algebra is a G^* module if we forget about the multiplicative structure.

We want to make the set of G^* modules and the set of G^* algebras into a category, so we must define what we mean by a morphism. So let A and B be G^* modules and

$$f : A \to B$$

a (continuous) linear map.

Definition 2.3.2 *We say that f is a **morphism** of G^* modules if for all $x \in A, a \in G, \xi \in g$ we have*

$$[\rho(a), f] = 0 \qquad (2.27)$$
$$[L_\xi, f] = 0 \qquad (2.28)$$
$$[\iota_\xi, f] = 0 \qquad (2.29)$$
$$[d, f] = 0. \qquad (2.30)$$

Notice that (2.28) is a consequence of (2.27) because of (2.23). If G is connected, (2.27) is a consequence of (2.28) for the same reason.

If, for all i,

$$f : A_i \to B_{i+k}$$

we say that f has **degree** k, with similar notation in the $(\mathbf{Z}/2\mathbf{Z})$-graded case. We say that a morphism of degree $\mathbf{k}$ is **even** if $\mathbf{k} = \mathbf{0}$ and **odd** if $\mathbf{k} = \mathbf{1}$.

If the morphism is even (especially if it is of degree zero which will frequently be the case) we could write conditions (2.27)–(2.30) as saying that $\forall a \in G, \xi \in g$,

$$\rho(a)f(x) = f(\rho(a)x)$$
$$L_\xi f(x) = f(L_\xi x)$$
$$\iota_\xi f(x) = f(\iota_\xi x)$$
$$df(a) = f(da).$$

Or, more informally, we could say that f preserves the G^* action.

It is clear that the composite of two G^* module morphisms is again a G^* module morphism, and hence that we have made the set of G^* module morphisms into a category.

We define a morphism between G^* algebras to be a map $f : A \to B$ which is an algebra homomorphism and satisfies (2.27)–(2.30). This makes the set of G^* algebras into a category.

We can make the analogous definitions for **Z**-graded G^* modules, algebras and morphisms.

If we have a G-action on a manifold, M, then $\Omega(M)$ is a G^* algebra in a canonical way. If M and N are G-manifolds and $F : M \to N$ is a G-equivariant smooth map, then the pullback map $F^* : \Omega(N) \to \Omega(N)$ is a morphism of G^* algebras. So the category of G^* algebras can be considered as an algebraic generalization of the category of G-manifolds. Our immediate task will be to translate various concepts from geometry to algebra:

2.3.1 Cohomology

By definition, the element d acts as a derivation of degree $+1$ with $d^2 = 0$ on A. So A is a cochain complex. We define $H(A) = H(A, d)$ to be the cohomology of A relative to the differential d. In case $A = \Omega(M)$ de Rham's theorem says that this is equal to $H^*(M)$.

Remarks.

1. $H^*(A)$ is a supervector space, and a superalgebra if A is. It is **Z**-graded if A is.

2. A morphism $f : A \to B$ induces a map $f_* : H^*(A) \to H^*(B)$ which is an algebra homomorphism in the algebra case. It is **Z**-graded in case we are in the category of **Z**-graded modules or algebras.

3. Condition (2.26) implies that $H^*(A)$ inherits the structure of a G-module. But notice that the connected component of the identity of G acts trivially. Indeed, if $\omega \in A$ satisfies $d\omega = 0$, then, for any $\xi \in g$ we have, by (2.5),

$$L_\xi \omega = d\iota_\xi \omega$$

so the cohomology class represented by $L_\xi \omega$ vanishes.

4. If $f : A \to B$ is a morphism, then the induced morphism

$$f_* : H^*(A) \to H^*(B)$$

is a morphism of G modules.

2.3.2 Acyclicity

If M is contractible, the de Rham complex $(\Omega(M), d)$ is acyclic, i.e., $A = \Omega(M)$ satisfies

$$H^k(A, d) = \begin{cases} \mathbf{F} & k = 0, \\ 0 & k \neq 0. \end{cases} \tag{2.31}$$

where $\mathbf{F}$ is the ground field, which is $\mathbf{C}$ in our case. We take this as the definition of acyclicity for a general A.

2.3.3 Chain Homotopies

Let A and B be two G^* modules. A linear map

$$Q : A \to B$$

is called a **chain homotopy** if it is odd, G-equivariant, and satisfies

$$\iota_\xi Q + Q \iota_\xi = 0 \quad \forall \xi \in g. \tag{2.32}$$

If A and B are $\mathbf{Z}$-graded (as we shall usually assume) we require that Q be of degree -1 in the $\mathbf{Z}$-gradation. The G-equivariance implies that

$$L_\xi Q - Q L_\xi = 0 \quad \forall \xi \in g. \tag{2.33}$$

Proposition 2.3.1 If $Q : A \to B$ is a chain homotopy then

$$\tau := dQ + Qd \tag{2.34}$$

is a morphism of G^* modules.

Proof. We have

$$d\tau = dQd = \tau d$$

and Q is assumed to be G-equivariant hence gives a g_0-morphism. We must check that $\iota_\xi \tau = \tau \iota_\xi \ \forall \xi \in g$. We have

$$
\begin{aligned}
\iota_\xi \tau &= \iota_\xi dQ + \iota_\xi Qd \\
&= \iota_\xi dQ - Q\iota_\xi d \\
&= -d\iota_\xi Q + L_\xi Q + Qd\iota_\xi - QL_\xi \\
&= (dQ + Qd)\iota_\xi \\
&= \tau \iota_\xi. \quad \square
\end{aligned}
$$

Let us redo the above argument in superlanguage: Since Q is odd, condition (2.32) says

$$[\iota_\xi, Q] = 0 \quad \forall \xi \in g$$

and the definition (2.34) can be written as

$$\tau := [d, Q]$$

and (2.33) as

$$[L_\xi, Q] = 0 \quad \forall \xi \in g.$$

By construction τ is an even G-morphism so $[L_\xi, \tau] = 0$ for all $\xi \in g$. Also

$$
\begin{aligned}
[\iota_\xi, \tau] &= [\iota_\xi, [d, Q]] \\
&= [[\iota_\xi, d], Q] - [d, [\iota_\xi, Q] \\
&= [L_\xi, Q] - 0 \\
&= 0,
\end{aligned}
$$

while

$$[d, \tau] = [d, [d, Q]] = [[d, d], Q] - [d, [d, Q]] = -[d, \tau] = 0. \quad \square$$

We say that two morphisms τ_0 and $\tau_1 : A \to B$ are **chain homotopic** and write

$$\tau_0 \simeq \tau_1$$

if there is a chain homotopy $Q : A \to B$ such that

$$\tau_1 - \tau_0 = Qd + dQ. \tag{2.35}$$

Notice that this implies that the induced maps on cohomology are equal:

$$\tau_0 \simeq \tau_1 \Rightarrow \tau_{0*} = \tau_{1*}. \tag{2.36}$$

We pause to remind the reader how chain homotopies arise in de-Rham theory: Suppose that

$$A = \Omega(Z), \quad B = \Omega(W)$$

where Z and W are smooth manifolds. Suppose that

$$\phi_0 : W \to Z, \quad \text{and} \quad \phi_1 : W \to Z$$

are smooth maps, and let

$$\tau_i := \phi_i^* : A \to B, \quad i = 0, 1.$$

We say that ϕ_0 and ϕ_1 are smoothly homotopic if there is a smooth map

$$\phi : W \times I \to Z$$

where I is the unit interval, and

$$\phi_0 = \phi(\cdot, 0), \quad \phi_1 = \phi(\cdot, 1).$$

We claim that this implies that τ_0 and τ_1 are chain homotopic.

Proof. For general $t \in I$, define

$$\phi_t : W \to Z, \quad \phi_t(w) = \phi(w, t)$$

and define

$$\xi_t : W \to TZ$$

by letting $\xi_t(w)$ be the tangent vector to the curve $s \mapsto \phi(w, s)$ at $s = t$. For $\sigma \in \Omega^{k+1}(Z)$ define

$$\phi_t^*(i(\xi_t)\sigma) \in \Omega^k(W)$$

by

$$\phi_t^*(i(\xi_t)\sigma)(\eta_1, \ldots, \eta_k) = \sigma(\xi_t(w), d\phi_t(\eta_1), \ldots, d\phi_t(\eta_k)), \quad \eta_1, \ldots \eta_k \in TW_w.$$

The "basic formula of differential calculus" asserts that

$$\frac{d}{dt}\phi_t^*\sigma = \phi_t^*(\iota(\xi_t)d\sigma + d\phi_t^*(\iota(\xi_t)\sigma).$$

(For a proof of a slightly more general formula, see [GS] page 158.) Define $Q : A \to B$ by

$$Q\sigma := \int_0^1 \phi_t^*(\iota(\xi_t)\sigma)dt.$$

Integrating the preceding equation from 0 to 1 shows that (2.35) holds. All the above is completely standard. Now suppose that Z and W are G-manifolds and that all the maps in question, ϕ_0, ϕ_1, ϕ, are G-equivariant. Then A and B are G^* modules, τ_0 and τ_1 are G^* morphism, and it follows from the above definition of Q that (2.32) holds, i.e. that Q is a chain homotopy.

Suppose that A and B are G^* algebras, and we are given an algebra homomorphism $\phi : A \to B$ which is a G^* morphism. We say that Q is a **chain homotopy relative to ϕ** or a **ϕ-homotopy** if, in addition to (2.32), Q satisfies the derivation identity

$$Q(xy) = Q(x)\phi(y) \pm \phi(x)Q(y), \quad \pm = (-1)^i, \quad \forall x \in A_i, \ y \in A. \quad (2.37)$$

This condition implies that Q is determined by its values on the generators of A. Conversely, suppose that we are given ϕ and a linear map $Q : A \to B$ satisfying (2.37) and which satisfies (2.32) on the generators. Then Q is a chain homotopy. Indeed, we must show that if (2.32) holds when evaluated at x and at y then it holds when evaluated on xy. We have, using (2.37),

$$\iota_\xi Q(xy) = [\iota_\xi Q(x)]\phi(y) \mp Q(x)\phi(\iota_\xi y) \pm [\iota_\xi \phi(x)]Q(y) + \phi(x)\iota_\xi Q(y)$$

since Q and ι_ξ are both odd. On the other hand

$$Q(\iota_\xi(xy)) = Q([\iota_\xi x]y \pm x\iota_\xi y) =$$

$$= Q(\iota_\xi x)\phi(y) \mp \phi(\iota_\xi x)Q(y) \pm Q(x)\phi(\iota_\xi y) + \phi(x)Q(\iota_\xi y),$$

and upon adding, the middle terms cancel.

2.3.4   Free Actions and the Condition (C)

There is no easy way in de Rham theory of detecting whether or not an action is free. But it is useful to weaken this condition to one that can be detected at the infinitesimal level:

Definition 2.3.3 *An action of G on M is said to be* **locally free** *if, the corresponding infinitesimal action of g is free, i.e., if, for every $\xi \neq 0 \in$ g, the vector field ξ_M^- generating the one parameter group $t \mapsto \exp{-t\xi}$ of transformations on M is nowhere vanishing.*

If the action is locally free, we can find linear differential forms, $\theta^1, \cdots, \theta^n$ on M which are everywhere dual to our basis $\xi_1, \ldots, \xi_n$ in the sense that

$$\iota_a \theta^b = \delta_a^b. \tag{2.38}$$

Conversely, if we have a G-action on a manifold on which there exist forms θ^a satisfying (2.38) then it is clear that the action is locally free.

A linear differential form ω is called **horizontal** if it satisfies

$$\iota_a \omega = 0 \quad a = 1, \ldots, n. \tag{2.39}$$

The local-freeness assumption says that the horizontal linear differential forms span a sub-bundle of the cotangent bundle, whose fiber at each point consists of covectors which vanish on the values of the vector fields coming from g. In other words, it says that the values of the vector fields coming from g form a vector sub-bundle of the tangent bundle, TM. The sub-bundle of T^*M spanned by the horizontal differential forms is called the **horizontal bundle**.

If the sub-bundle spanned by the forms satisfying (2.38) is G-invariant, then the forms θ^i are usually called **connection forms**; at least this is the standard terminology when the G-action makes M into a principal bundle over some base B (so that the action is free and not just locally free). In the standard terminology, one usually considers a "connection form" to be a g-valued one form $\theta \in \Omega^1(M) \otimes$ g. Relative to our chosen basis of g, $\theta = \theta^i \otimes \xi_i$ where the θ^i are the connection forms defined above.

Suppose we have a locally free action of G on M, and we put a Riemann metric on M. This splits the cotangent bundle into a subbundle C complementary to the horizontal bundle whose fiber at each point is isomorphic to g^*. Hence our basis of g picks out a dual basis of the fiber of C at each point, i.e. a set of linear differential forms satisfying (2.38). In general, the sub-bundle C will not be G-invariant. But if the group G is compact, we can choose our Riemann metric to be G-invariant by averaging over the group, in which case the sub-bundle C will also be G-invariant.

Since $\iota_b \theta^j$ is constant, we have

$$
\begin{aligned}
0 &= L_a \iota_b \theta^j \\
&= ([L_a, \iota_b] + \iota_b L_a) \theta^j \\
&= c_{ab}^k \iota_k \theta^j + \iota_b L_a \theta^j \\
&= c_{ab}^j + \iota_b L_a \theta^j \quad \text{so} \\
L_a \theta^j &= -c_{ab}^j \theta^b + \omega_a^j
\end{aligned}
$$

where ω_a^j is horizontal, i.e. satisfies (2.39).

If the sub-bundle C is G-invariant, then all the $\omega_a^j = 0$ and we get

$$L_a \theta^j = -c_{ab}^j \theta^b. \tag{2.40}$$

Abstracting from these properties, we make the following definition:

Definition 2.3.4 *A G^* algebra A is said to be of type (C) if there are elements $\theta^i \in A_1$ (called **connection elements**) which satisfy (2.38), and such that the subspace $C \subset A_1$ that they span is invariant under G.*

If G is connected, condition (2.40) implies that the space spanned by the θ^i is G-invariant. So if G is connected then being of type (C) amounts to the existence of θ^i satisfying (2.38) and (2.40).

Usually the properties of A that we will study will be independent of the specific choice of the connection elements, θ^i. This is in analogy to the geometrical case where the topological properties of a principal bundle are independent of the choice of connection.

It follows from (2.38) and (2.40) that

$$d\theta^a = -\frac{1}{2} c_{ij}^a \theta^i \theta^j + \mu^a \tag{2.41}$$

where the μ^a are two-forms satisfying

$$\iota_a \mu^b = 0. \tag{2.42}$$

In the case of principal bundles and connection forms, the forms μ^a are called the **curvature forms** associated to the given connection. For general algebras of type (C) we will call the elements μ^a occurring in (2.41) the **curvature elements** corresponding to the connection elements $\{\theta^a\}$.

If we are given $\theta^i \in A_1$ satisfying (2.38) and (2.40) then, as we have seen, (2.41) and (2.42) are consequences. If we apply d to (2.41) we find (using Jacobi's identity) that

$$d\mu^a = -c^a_{ij}\theta^i\mu^j \tag{2.43}$$

and from this equation and from (2.5) and (2.42) that

$$L_i\mu^a = -c^a_{ij}\mu^j. \tag{2.44}$$

If A is any G^* algebra and B is a G^* algebra of type (C), with connection elements θ^i_B then $A \otimes B$ is again an algebra of type (C) with connection elements $1 \otimes \theta^i_B$.

Let us return to conditions (2.38) and (2.40). Consider the map $C : g^* \to A_1$, given by

$$C(x^i) = \theta^i$$

where $x^1, \ldots, x^n$ is the basis of g^* dual to the basis $\xi_1, \ldots, \xi_n$ that we have chosen of g. Thus the subspace, C, spanned by the θ^i is just the image of C,

$$C = C(g^*).$$

Condition (2.38) is then equivalent to

$$\iota_\xi\left(C(\theta)\right) = \langle\theta,\xi\rangle, \quad \forall \xi \in g, \ \theta \in g^*. \tag{2.45}$$

Notice that if C satisfies this equation, so does

$$a \circ C \circ \mathrm{Ad}^\sharp_{a^{-1}}$$

where $\mathrm{Ad}^\sharp$ denotes the co-adjoint representation, the representation of G on g^* contragredient to the adjoint representation:

$$\langle\mathrm{Ad}^\sharp_b\theta,\xi\rangle := \langle\theta,\mathrm{Ad}_{b^{-1}}\xi\rangle.$$

Indeed,

$$\begin{aligned}
\iota_\xi\left(aC(\mathrm{Ad}^\sharp_{a^{-1}}\theta)\right) &= a\iota_{\mathrm{Ad}_{a^{-1}}\xi}C(\mathrm{Ad}^\sharp_{a^{-1}}\theta) \quad \text{by (2.25)} \\
&= \langle\mathrm{Ad}^\sharp_{a^{-1}}\theta,\mathrm{Ad}_{a^{-1}}\xi\rangle \\
&= \langle\theta,\xi\rangle.
\end{aligned}$$

(In passing from the first to the second line we are making the mild assumption that G acts trivially on the scalars, considered as a one-dimensional subspace of A_0. This is usually what is meant when we talk an automorphism of an algebra with unit — that the automorphism preserve the unit.)

The condition that C be invariant is the same as the condition that C be equivariant, i.e. that

$$a \circ C \circ \mathrm{Ad}^\sharp_{a^{-1}} = C \quad \forall a \in G.$$

If G is compact, and we are given a C satisfying (2.45), then averaging $C_a :=$ $a \circ C \circ \mathrm{Ad}^{\sharp}_{a^{-1}}$ over the group, i.e. considering the integral

$$\int_G C_a \, da$$

with respect to Haar measure gives a new C which is equivariant. So in the case that G is compact, a G^* algebra is of type (C) if and only if there exist elements satisfying (2.38).

2.3.5   The Basic Subcomplex

If the action of G on M is free and G is compact, the quotient space $X = M/G$ is a manifold and the projection

$$\pi : M \to X$$

is a principal G-fibration. The subcomplex

$$\pi^* \Omega(X) \subset \Omega(M)$$

is called the complex of **basic forms** since they are images of forms coming from the base X under the injective map π^*. Since π^* is injective, the complex of basic forms is isomorphic to $\Omega(X)$. It is easy to detect when a form is basic: ω is basic if and only if it is G-invariant and horizontal, i.e. satisfies (2.39). Moreover, if G is connected, being G-invariant is equivalent to satisfying

$$L_a \omega = 0, \quad a = 1, \dots, n. \tag{2.46}$$

For an arbitrary G^* module A we define A_{bas} to be the set of all elements which are G-invariant and satisfy (2.39). If G is connected we can replace G-invariance by (2.46). The set of elements of A_{bas} are called **basic**. It follows from (2.5)

$$dA_{\mathrm{bas}} \subset A_{\mathrm{bas}},$$

in other words A_{bas} is a subcomplex of A. We will call its cohomology the **basic cohomology** of A. We will denote this basic cohomology by

$$H(A_{\mathrm{bas}}, d)$$

or, more simply, by

$$H_{\mathrm{bas}}(A).$$

By definition, $\rho(a)$, $a \in G$, L_ξ and ι_ξ, $\xi \in \mathfrak{g}$ all act trivially on A_{bas}. So A_{bas} is a G^* submodule of A, but the only non-trivial action is that of d.

In the case that A is a G^* algebra, it follows from the fact that G acts as automorphisms and g_{-1} as derivations that A_{bas} is a G^* subalgebra. In this case, H_{bas} inherits an algebra structure.

Let $\phi : A \to B$ be a morphism of G^* modules. It follows immediately from the definitions that

$$\phi(A_{\text{bas}}) \subset B_{\text{bas}}$$

and hence that ϕ induces a linear map

$$\phi_\flat : H_{\text{bas}}(A) \to H_{\text{bas}}(B).$$

In case ϕ is a homomorphism (and morphism) of G^* algebras, the induced map $\phi_\flat$ is an algebra homomorphism.

2.4   Equivariant Cohomology of G^* Algebras

Let E be a G^* algebra which is acyclic and satisfies condition (C). Given any G^* algebra A we will define its equivariant cohomology ring $H_G(A)$ to be the cohomology ring of the basic subcomplex of $A \otimes E$:

$$H_G(A) := H_{\text{bas}}(A \otimes E) = H\left((A \otimes E)_{\text{bas}}, d\right). \tag{2.47}$$

We make the same definition (without the algebra structure) in the case of G^* modules.

Notice that this definition mimics the definition (1.3) in the framework of G^* algebras: we have replaced the space $M \times E$ where E is a classifying space (a free, acyclic G-space) by $\Omega(M) \otimes E$ where E is an acyclic G^* algebra of type (C), and then $\Omega(M)$ by a general G^* module A. We have replaced the cohomology of the quotient by the basic cohomology.

To show that the definition (2.47) is legitimate we will have to address the same issues we faced in Chapter 1: Does such an E exist and is the definition independent of the choice of E? We will postpone the independence question until Section 4.4. In the next section we will construct a rather complicated acyclic G^* algebra satisfying condition (C), but one which is closely related to the geometric construction in Chapter 1. We will use it to prove that the equivariant cohomology of a manifold M (as a topological space) is the same as the equivariant cohomology of $\Omega(M)$ (as a G^* module). In the next chapter we will introduce the Weil algebra which is the most economical choice of acyclic G^* algebra satisfying condition (C); most economical in a sense that we will make precise.

Let us continue to assume that the definition (2.47) is legitimate. If $\phi : A \to B$ is a morphism of G^* modules, we may choose the same E to compute the equivariant cohomology of both A and B. Then

$$\phi \otimes \text{id} \; : \; A \otimes E \to B \otimes E$$

is a morphism of G^* modules and we may try to define

$$\phi_G : H_G(A) \to H_G(B)$$

as

$$\phi_G := (\phi \otimes \mathrm{id}\,)_b.$$

The proof that we will give of the legitimacy of (2.47) will also show that this definition is independent of the choice of E. It then follows that

$$A \mapsto H_G(A), \quad \phi \mapsto \phi_G$$

is a functor. We leave the proof of the following as an easy exercise for the reader:

Proposition 2.4.1 *If two G^* morphisms, $\phi_i : A \to B$, $i = 0, 1$ are chain homotopic, then*

$$(\phi_0)_G = (\phi_1)_G.$$

2.5   The Equivariant de Rham Theorem

Theorem 2.5.1 *Let G be a compact Lie group acting on a smooth compact manifold M. then*

$$H^*_G(M) = H_G\left(\Omega(M)\right). \tag{2.48}$$

Without loss of generality we can assume that G is a closed subgroup of $U(n)$. Let $\mathbf{C}^\infty$ denote the space of all sequences $(z_1, z_2 \ldots, z_n, \ldots)$ with $z_i = 0$ for i sufficiently large. So $\mathbf{C}^\infty = \bigcup \mathbf{C}^k$ where $\mathbf{C}^k$ consists of all sequences with $z_i = 0$, $i > k$.

Let

$$\mathcal{E} = \mathcal{E}^{(n)}$$

denote the set of all orthonormal n-tuples $\mathbf{v} = (v_1, \ldots, v_n)$ with $v_i \in \mathbf{C}^\infty$. For $k > n$ let $\mathcal{E}_k$ be the set of all orthonormal n-tuples with $v_i \in \mathbf{C}^k$. From the inclusion

$$\mathbf{C}^k \to \mathbf{C}^{k+1}, \quad (z_1, \ldots, z_k) \mapsto (z_1, \ldots, z_k, 0)$$

one gets inclusions

$$i_k : \mathcal{E}_k \to \mathcal{E}_{k+1}$$

and, by composing them, inclusions

$$j_{k,m} : \mathcal{E}_k \to \mathcal{E}_m, \quad m > k$$

which compose consistently and give rise to inclusions

$$j_k : \mathcal{E}_k \to \mathcal{E}.$$

We use these inclusions to put the "final topology" on $\mathcal{E}$ (using the terminology of Bourbaki, [Bour] I-2: A set $U \subset \mathcal{E}$ is declared to be open if and only if each of the subsets $j_k^{-1}(U)$ is open.) As a consequence, a series of points converges if and only if there exists some k such that all the points lie in $\mathcal{E}_k$ and the sequence converges there. In particular, a continuous map, f of a compact space X into $\mathcal{E}$ satisfies $f(X) \subset \mathcal{E}_k$ for some k.

Proposition 2.5.1 *Let X be a compact m-dimensional manifold. Every continuous map $f : X \to \mathcal{E}$ is contractible to a point.*

Proof. We know that $f(X) \subset \mathcal{E}_k$ for some k. To prove the proposition it suffices to prove

Proposition 2.5.2 *A continuous map $f : X \to \mathcal{E}_k = \mathcal{E}_k^{(n)}$ is contractible to a point if $k \geq m + n$.*

Proof (by induction on n): Consider the fibration

$$\gamma : \mathcal{E}_k^{(n)} \to \mathcal{E}_k^{(n-1)}, \quad (v_1, \dots, v_n) \mapsto (v_1, \dots, v_{n-1}).$$

By induction, $\gamma \circ f$ is contractible to a point p_0 and hence, by the covering homotopy property (see [BT] page 199) f is homotopic to a map $h : X \to \mathcal{E}_k^{(n)}$ whose image sits in the fiber over p_0. But this fiber is a sphere $S^{2(k-(n-1))-1}$, so if $k \geq n + m$ this map is contractible to a point. $\square$

Let $\Omega(\mathcal{E})$ be the inverse limit of the sequence of projections

$$\cdots \leftarrow \Omega(\mathcal{E}_k) \leftarrow \Omega(\mathcal{E}_{k+1}) \leftarrow \Omega(\mathcal{E}_{k+2}) \leftarrow \cdots.$$

Proposition 2.5.3 *$\Omega(\mathcal{E})$ is acyclic.*

Proof. By Proposition 2.5.2, (and using, say singular homology and cohomology)

$$H^\ell(\mathcal{E}_k) = 0, \quad \ell > 0$$

if $k \gg \ell$ is sufficiently large. Thus, by the usual de Rham theorem, if $\mu \in \Omega^\ell(\mathcal{E})$ is closed, then

$$\mu_k := j_k^* \mu = d\nu_k$$

for k sufficiently large. We claim that we can choose the ν_k consistently, i.e. such that

$$\nu_k = i_k^* \nu_{k+1}.$$

Indeed for any choice of ν_{k+1} we have

$$i_k^* \nu_{k+1} - \nu_k = d\lambda.$$

Choose an $(\ell - 2)$-form β on $\mathcal{E}_{k+1}$ such that $i_k^* \beta = \lambda$. Replacing ν_{k+1} by $\nu_{k+1} - d\beta$ gives us a consistent choice, and proceeding inductively we get a consistent choice for all large k. Hence we can find a $\nu \in \Omega(\mathcal{E})$ with $j_k^* \nu = \nu_k$ and $d\nu = \mu$. $\square$

$U(n)$ acts freely on $\mathcal{E}$ by the action (1.9), and this induces an action of $U(n)$ on $\Omega(\mathcal{E})$. So we can apply (1.3) to conclude that

$$H_G^*(M) = H^* \left((M \times \mathcal{E})/G \right). \tag{2.49}$$

Proposition 2.5.4 *$\Omega(\mathcal{E})$ satisfies property (C).*

Proof. Let z_{ij} be the functions defined on $\mathcal{E}$ by setting $z_{ij}(\mathbf{v}) = i$th coordinate of the vector v_j where $\mathbf{v} = (v_1, \ldots, v_n)$ and let Z be the matrix with entries z_{ij}. So Z has only finitely many non-zero entries when evaluated on any $\mathcal{E}_k$. We may thus form the matrix

$$\Theta := Z^t d\overline{Z}$$

which is an element of $\Omega(\mathcal{E})$ from whose components we get θ's with the property (C).

Let $\Omega(M \times \mathcal{E})$ be the inverse limit of the sequence

$$\cdots \leftarrow \Omega(M \times \mathcal{E}_k) \leftarrow \Omega(M \times \mathcal{E}_{k+1}) \leftarrow \cdots.$$

We claim that

$$H^* \left((M \times \mathcal{E})/G \right) = H_{\mathrm{bas}} \left(\Omega(M \times \mathcal{E}) \right). \tag{2.50}$$

Proof. This follows from the fact that for each i the two sequences

$$\cdots \leftarrow H^i \left((M \times \mathcal{E}_k)/G \right) \leftarrow H^i \left((M \times \mathcal{E}_{k+1})/G \right) \leftarrow \cdots$$

and

$$\cdots \leftarrow H^i \left((\Omega(M \times \mathcal{E}_k)_{\mathrm{bas}} \right) \leftarrow H \left((\Omega(M \times \mathcal{E}_{k+1})_{\mathrm{bas}} \right) \leftarrow \cdots$$

are termwise isomorphic. □

Proposition 2.5.5 *The inclusion map*

$$\Omega(M) \otimes \Omega(\mathcal{E}) \to \Omega(M \times \mathcal{E})$$

induces an isomorphism on cohomology:

$$H \left((\Omega(M) \otimes \Omega(\mathcal{E}))_{\mathrm{bas}} \right) \to H \left((\Omega(M \times \mathcal{E})_{\mathrm{bas}} \right).$$

Proof. By a spectral sequence argument (see Theorem 6.7.1) it is enough to see that the inclusion induces an isomorphism

$$H(\Omega(M) \otimes \Omega(\mathcal{E})) \to H(\Omega(M \times \mathcal{E}))$$

on ordinary cohomology. But the acyclicity of $\Omega(\mathcal{E})$ and the contractibility of $\mathcal{E}$ imply that this map is just the identity map of $H^*(M)$ into itself. □

Since $\Omega(\mathcal{E})$ is a G^* algebra which is acyclic and has property (C), we conclude that

$$H_G(\Omega(M)) = H \left(\Omega(M \times \mathcal{E})_{\mathrm{bas}} \right)$$

and hence that

$$H_G(\Omega(M)) = H_G^*(M).$$

2.6   Bibliographical Notes for Chapter 2

1. For a more detailed exposition of the super ideas discussed in section 2.2 see Berezin [Be], Kostant [Ko1] or Quillen [Qu].

2. The term "G^* module" is due to us; however the notion of G^* module is due to Cartan. (See "Notions d'algèbre différentielles, ..." page 20, lines 15-20.)

3. In this monograph the two most important examples which we will encounter of G^* modules are the de Rham complex, $\Omega(M)$, and the equivariant de Rham complex, $\Omega_K(M)$ (which we'll encounter in Chapter 4. If M is a $(G \times K)$-manifold this complex is a G^* module). From these two examples one gets many refinements: e.g., the complex of compactly supported de Rham forms, the complex of de Rham currents, the relative de Rham complex associated with a G-mapping $f : X \longrightarrow Y$ (see [BT] page 78), inverse and direct limits of de Rham complexes (an example of which is the complex $\Omega(\mathcal{E})$, in §2.5), the Weil complex (see Chapter 3), the Mathai-Quillen complex (see Chapter 7), the universal enveloping algebra of the Lie superalgebra, $\tilde{g}, \cdots$.

4. The subalgebra $g_{-1} \oplus g_0$ of $\tilde{g}$ is the tensor product of the Lie algebra g and the commutative superalgebra, $\mathbf{C}[x]$, generated by an element, x, of degree -1. The representation theory of the Lie superalgebra

$$g \otimes \mathbf{C}[x_1, \ldots, x_n],$$

with generators $x_1, \cdots, x_n$ of arbitrary degree, has been studied in detail by Cheng, (See [Ch]).

5. Another interesting representation of a Lie superalgebra on $\Omega(M)$ occurs in some recent work of Olivier Mathieu: Let M be a compact symplectic manifold of dimension $2n$ with symplectic form ω. Since ω is a non-degenerate bilinear form on the tangent bundle of M, it can be used to define a Hodge star operator

$$\star : \Omega^k(M) \longrightarrow \Omega^{2n-k}(M).$$

Let

$$E : \Omega^k \longrightarrow \Omega^{k+2}$$

be the operator, $E\mu = \omega \wedge \mu$. Let

$$F : \Omega^k \longrightarrow \Omega^{k-2}$$

be the operator $- \star E \star$. Let

$$\delta : \Omega^k \longrightarrow \Omega^{k-1}$$

be the operator $(-1)^k * d*$ and let

$$A : \Omega^k \longrightarrow \Omega^k$$

be the operator $(n-k)\cdot$ identity. These operators define a representation on $\Omega(M)$ of the simple five-dimensional Lie superalgebra

$$g^{-2} \oplus g^{-1} \oplus g^0 \oplus g^1 \oplus g^2$$

with generators, $F \in g^{-2}$, $\delta \in g^{-1}$, $H \in g^0$, $d \in g^1$ and $E \in g^2$, and relations: $[E,F] = H$, $[H,F] = -2F$, $[H,E] = 2E$, $[F,d] = \delta$ and $[E,\delta] = -d$. From the existence of this representation Mathieu [Mat] deduces some fascinating facts about symplectic Hodge theory: For instance M is said to satisfy the *Brylinski condition* if every cohomology class admits a harmonic representative. Mathieu proves that M satisfies this condition if and only if the strong Lefshetz theorem holds: i.e., iff the map

$$E^k : H^{n-k}(M) \longrightarrow H^{n+k}(M)$$

is bijective. See [Mat].

6. The fact that, in equivariant de Rham theory, there is no way to differentiate between *free* G-actions and actions which are only *locally free* has positive, as well as negative, implications: The class of manifolds for which $\Omega(M)$ satisfies condition (C) includes not only principal G-bundles but many other interesting examples besides (for instance, in symplectic geometry, the non-critical level sets of moment mappings!)

7. Atiyah and Bott sketch an alternative proof of the equivariant de Rham theorem in Section 4 of [AB]. One of the basic ingredients in their proof is the "Weil model" of which we will have much more to say in the next two chapters.

Chapter 3

The Weil Algebra

3.1 The Koszul Complex

Let V be an n-dimensional vector space, and let $\wedge = \wedge(V)$ be the exterior algebra of V considered as a commutative superalgebra, and let $S = S(V)$ be the symmetric algebra considered as an algebra all of whose elements are even. So we assign to each element of $\wedge V$ its exterior degree, but each element of $S^k(V)$ is assigned the degree $2k$. The **Koszul algebra** is the tensor product $\wedge \otimes S$.

The elements $x \otimes 1 \in \wedge^1 \otimes S^0$ and $1 \otimes x \in \wedge^0 \otimes S^1$ generate $\wedge \otimes S$. The **Koszul operator** d_K is defined as the derivation extending the operator on generators given by

$$d_K(x \otimes 1) = 1 \otimes x, \quad d_K(1 \otimes x) = 0.$$

Clearly $d_K^2 = 0$ on generators, and hence everywhere, since d_K^2 is a derivation. We can also use this same argument, and the fact that the commutator of two odd derivations is an even derivation, to prove that the Koszul operator is acyclic. Indeed, let Q be the derivation defined on generators by

$$Q(x \otimes 1) = 0, \quad Q(1 \otimes x) = x \otimes 1.$$

So $Q^2 = 0$ and $[Q, d_K] = \mathrm{id}$ on generators. But since $[Q, d_K]$ is an even derivation, we conclude that

$$[Q, d_K] = (k + \ell)\,\mathrm{id} \qquad \text{on } \wedge^k \otimes S^\ell.$$

Thus the only cohomology of d_K lies in $\wedge^0 \otimes S^0$, which is the field of scalars. In fact, the cohomology is the field of scalars, since $d_K 1 = 0$.

It will be convenient for us to write all of this in terms of a basis. Let $x^1, \ldots, x^n$ be a basis of V and define

$$\theta^i := x^i \otimes 1 \tag{3.1}$$

and

$$z^i := 1 \otimes x^i. \tag{3.2}$$

Then the Koszul operator $d = d_K$ is expressed in terms of these generators as

$$d\theta^i = z^i \tag{3.3}$$

and

$$dz^i = 0. \tag{3.4}$$

and the operator Q is given by

$$Qz^i = \theta^i, \quad Q\theta^i = 0.$$

3.2   The Weil Algebra

The Weil algebra is just the Koszul algebra of g^*:

$$W := \wedge(g^*) \otimes S(g^*). \tag{3.5}$$

The group G acts on g via the adjoint representation, hence acts on g^* via the contragredient to the adjoint representation (the coadjoint representation) and hence acts as superalgebra automorphisms on W. A choice of basis, $\xi_1, \ldots, \xi_n$ of g induces a dual basis of g^* and hence generators, $\theta^1, \ldots, \theta^n, z^1, \ldots z^n$ which satisfy

$$L_a \theta^b = -c^b_{ak} \theta^k \tag{3.6}$$

and

$$L_a z^b = -c^b_{ak} z^k. \tag{3.7}$$

The Koszul differential $d = d_K$ is clearly G-equivariant. This means that we have an action of the $g_0 \oplus g_1$ part of $\tilde{g}$ on W. We would like to define the action of g_{-1}, i.e. prescribe the operators ι_a, so as to get a G^* action which is acyclic and has property (C) with the θ^b as connection elements. Recall that if G is connected, as we shall assume, property (C) means that, in addition to (3.6), the elements θ^i satisfy (2.38). So we *define* the action of ι_a on the θ^b by (2.38), i.e.

$$\iota_a \theta^b = \delta^b_a.$$

Since $d\iota_a + \iota_a d = L_a$ we are forced to have

$$\iota_a z^b = \iota_a d\theta^b = (\iota_a d + d\iota_a)\theta^b = L_a \theta^b$$

so

$$\iota_a z^b = -c^b_{ak} \theta^k. \tag{3.8}$$

So we use (2.38) and (3.8) to define the action of ι_a of generators, and extend as derivations to all of W. To check that we get an action of $\tilde{g}$ on W, we need only check that the conditions (2.17)–(2.22) hold on generators, which we have arranged to be true. We have proved:

Theorem 3.2.1 *W is an acyclic G^* algebra satisfying condition (C).*

We recall that W_{hor} is defined to be the set of all elements in W satisfying (2.39). We claim that

$$W = \wedge(g^*) \otimes W_{\mathrm{hor}}. \tag{3.9}$$

Proof of (3.9). Define

$$\mu^b := z^b + \frac{1}{2} c^b_{jk} \theta^j \theta^k. \tag{3.10}$$

Then

$$
\begin{aligned}
\iota_a \mu^b &= \iota_a z^b + \frac{1}{2} c^b_{jk} \delta^j_a \theta^k - \frac{1}{2} c^b_{jk} \theta^j \delta^k_a \\
&= \iota_a z^b + c^b_{a\ell} \theta^\ell \\
&= 0
\end{aligned}
$$

by (3.8). So the μ^b are horizontal elements of W. Moreover,

$$z^b = \mu^b - \frac{1}{2} c^b_{jk} \theta^j \theta^k$$

so we can use the θ^a and μ^b as generators of W. So W is the tensor product of the exterior algebra in the θ and the polynomial ring $\mathbf{C}[\mu^1, \ldots, \mu^n]$, and it is clear that an element in this decomposition is horizontal if and only if it lies in $\mathbf{C}[\mu^1, \ldots, \mu^n]$, i.e.

$$W_{\mathrm{hor}} = \mathbf{C}[\mu^1, \ldots, \mu^n]. \tag{3.11}$$

This completes the proof of (3.9)

Identifying $\mathbf{C}[\mu^1, \ldots, \mu^n]$ with the polynomial ring $S(g^*)$ and recalling that W_{bas} are the G-invariant elements of W_{hor} we obtain:

$$W_{\mathrm{bas}} = S(g^*)^G. \tag{3.12}$$

We obtained the element μ^b by adding the term $\frac{1}{2} c^b_{jk} \theta^j \theta^k$ to z^b. From the definition of the structure constants, the element

$$c^b_{jk} \xi_b \theta^j \otimes \theta^k$$

is precisely the map of $g \otimes g \to g$ given by Lie bracket. Hence the (old-fashioned) Jacobi identity for g can be expressed as

$$L_a \left(\frac{1}{2} c^b_{ij} \theta^i \theta^j \right) = -c^b_{ak} \left(\frac{1}{2} c^k_{ij} \theta^i \theta^j \right).$$

Therefore

$$L_a \mu^b = -c^b_{ak} \mu^k. \tag{3.13}$$

We will now show that the operator d acts trivially on W_{bas}. For this purpose, we first compute $d\mu^a$: We have, by (3.3) and (3.10),

$$d\theta^a = -\frac{1}{2}c^a_{ij}\theta^i\theta^j + \mu^a. \tag{3.14}$$

If we apply d we obtain

$$
\begin{aligned}
d\mu^b &= \frac{1}{2}c^b_{ij}d\theta^i\theta^j - \frac{1}{2}c^b_{ij}\theta^i d\theta^j \\
&= -c^b_{ij}\theta^i\mu^j + \cdots \quad (*)
\end{aligned}
$$

where the remaining terms cancel by Jacobi's identity. Thus

$$d\mu^a = -c^a_{ij}\theta^i\mu^j. \tag{3.15}$$

Combining (3.15) with (3.13) we can rewrite (3.15) as

$$d\mu^a = (\theta^b L_b)\mu^a.$$

Now a derivation followed by a multiplication is again a derivation, so the operator $\theta^b L_b$ is (an odd) derivation as is d. Since the μ^a generate W_{hor} we conclude that

$$dw = \theta^b L_b w \quad \forall\, w \in W_{\mathrm{hor}}. \tag{3.16}$$

In particular, if $w \in W_{\mathrm{bas}} = (W_{\mathrm{hor}})^G$, $L_a w = 0$, $a = 1,\dots,n$. Hence

$$dw = 0 \quad \forall w \in W_{\mathrm{bas}}. \tag{3.17}$$

To summarize, we have proved:

Theorem 3.2.2 *The basic cohomology ring of W is $S(g^*)^G$.*

Equation (3.14) is known as the **Cartan structure equation** and equation (3.15) is known as the **Bianchi identity** for the Weil algebra.

In the usual treatments of the Weil algebra, (3.9) is taken as the definition of the Weil algebra, where W_{hor} is defined as in (3.11), that is

$$W_{\mathrm{hor}} = \mathbf{C}[\mu^1,\dots,\mu^n] = S(g^*).$$

Of course, the $S(g^*)$ occurring in this version is different from our original $S(g^*)$; it is obtained from it by the supersymmetric "change of variables" (3.10). With the generators θ^a and μ^a, the action of g_{-1} is defined to be $\iota_a\theta^b = \delta^b_a$ and $\iota_a\mu^b = 0$, and the action of g_0 is defined by (3.6) on the θ^a and by (3.13) on the μ^a The linear space spanned by the θ^a is just a copy of g^* (which generates the subalgebra $\wedge(g^*)$) and the linear space spanned by the μ^a is a copy of g^* which generates $W_{\mathrm{hor}} \cong S(g^*)$. Both (3.6) and (3.13) describe the standard, coadjoint, action of g on g^*. The action of d in the standard treatments is defined to be (3.14) and (3.15) on the generators θ^a

and μ^a (and extended so as to be a derivation). One must do some work to then prove that $d^2 = 0$ and that the Weil algebra is acyclic. The advantage of using supersymmetric methods such as the "change of variables" (3.10) is that these facts are immediate consequences of the existence and acyclicity of the Koszul algebra. We will see another illustration of the power of this technique when we come to the Mathai-Quillen isomorphism in Chapter 4.

There *is* an important interpretation of the operator d which is natural from the point of view of the standard treatment. (We will not use any of the following discussion in the rest of the book): We may think of $W_{\text{hor}} = S(g^*)$ as a g-algebra, that is as an algebra which is a g-module with g acting as derivations. Then we may use the Chevalley-Eilenberg prescription for computing the Lie algebra cohomology of $S(g^*)$ where the complex is taken to be $\wedge(g^*) \otimes S(g^*)$ with differential operator d_{CE} given on generators by

$$d_{CE}\theta^a := -\frac{1}{2}c_{ij}^a\theta^i\theta^j, \quad d_{CE}\mu^a := -c_{ij}^a\theta^a\mu^b.$$

Then (3.14) and (3.15) say that

$$d = d_{CE} + d_K$$

where
$$d_K\theta^a := \mu^a, \quad d_K\mu^a = 0.$$

So we may think of $\wedge(g^*) \otimes W_{\text{hor}}$ as a copy of the Koszul complex with operator d_K. The net effect of the supersymmetric change of variables (3.10) going from our original $\wedge(g^*) \otimes S(g^*)$ to $\wedge(g^*) \otimes W_{\text{hor}}$ is to introduce Lie algebra cohomology into the picture by adding the Chevalley-Eilenberg operator.

3.3 Classifying Maps

In this section we wish to establish the algebraic analogue of Proposition 1.1.1. Recall that this proposition asserts that if G acts freely on a topological space X, and if E is a classifying space for G then there exists a G-equivariant map $h : X \to E$ (uniquely constructed up to homotopy) and hence a canonical map

$$f^* : H(E/G) \to H^*(X/G).$$

In our algebraic analogue (where arrows are reversed) W will play the role of the classifying space:

Let A be a G^* algebra of type (C). We claim that

Theorem 3.3.1 *There exists a G^* algebra homomorphism $\rho : W \to A$. Any two such are chain homotopic.*

Proof. If we choose $\theta_A^a \in A_1$ satisfying (2.38) and (2.40), then the map

$$\rho(\theta^a) = \theta_A^a$$

extends uniquely to a G^* homomorphism, since $W(g)$ is freely generated as an algebra by the θ^a and $d\theta^a$. This establishes the existence. If ρ_0 and ρ_1 are two such homomorphisms, define ρ_t, $0 \leq t \leq 1$ by first defining $\rho_t : W_1 \to A$ by

$$\rho_t(\theta^a) = (1-t)\rho_0(\theta^a) + t\rho_1(\theta^a)$$

i.e.

$$\rho_t = (1-t)\rho_0 + t\rho_1 \quad \text{on } W^1.$$

This map satisfies $\rho_t \iota_\xi = \iota_\xi \rho_t$ and is G-equivariant. So it extends to a G^* algebra homomorphism which we shall also denote by ρ_t. Let Q_t be the ρ_t chain homotopy defined by

$$Q_t \theta^a = 0, \quad Q_t d\theta^a = \frac{d}{dt}\rho_t(\theta^a)$$

on our generators, and extended by (2.37) (with $\phi = \rho_t$). It clearly satisfies (2.32) on these generators, and so is a chain homotopy relative to ρ_t. Then

$$Q := \int_0^1 Q_t dt$$

is the desired chain homotopy between ρ_0 and ρ_1. □

Since ρ is a G^* morphism, it maps W_{bas} into A_{bas} and hence the basic cohomology ring of W into the basic cohomology ring of A. Moreover, since ρ is unique up to chain homotopy, this map does not depend on ρ. Hence, by Theorem 3.2.2, we have proved

Theorem 3.3.2 *There is a canonical map*

$$\kappa_G : S(g^*)^G \to H_{\text{bas}}(A).$$

We shall call this map the **Chern-Weil map** or **characteristic homomorphism**. For a slightly different version of it see Section 4.5.

Suppose we have chosen the "connection elements" $\theta_A^a \in A$ satisfying (2.38) and hence the homomorphism $\rho : W \to A$ of the Weil algebra into A with $\rho(\theta^a) = \theta_A^a$. Define

$$\mu_A^a := \rho(\mu^a).$$

Since ρ is a G^* morphism, (3.14) and (3.15) imply

$$d\theta_A^a = -\frac{1}{2}c_{ij}^a \theta_A^i \theta_A^j + \mu_A^a \tag{3.18}$$

$$d\mu_A^a = -c_{ij}^a \theta_A^i \mu_A^j. \tag{3.19}$$

These are known as the Cartan equations and the Bianchi identity, or more simply as the Cartan structure equations for A. We could have derived them directly from the defining equations (2.38):

$$\iota_a \theta_A^b = \delta_a^b$$

and
$$L_a\theta_A^b = -c_{aj}^b\theta_A^j.$$
Indeed, since $L_a\theta_A^b = [d, \iota_a]\theta^b = \iota_a d\theta^b$ and

$$\iota_a\left(-\frac{1}{2}c_{ij}^b\theta_A^i\theta_A^j\right) = -c_{aj}^b\theta^j$$

we conclude that $d\theta_A^a$ differs from $-\frac{1}{2}c_{ij}^a\theta_A^i\theta_A^j$ by an element of degree two which is horizontal, and which we could define as μ_A^a and so get (3.18). Equation (3.19) then follows from (3.18) by applying d and using the Jacobi identity as we did for the case of the Weil algebra.

Theorem 3.3.1 can be thought of as saying that the Weil algebra is the simplest $G^\star$ algebra satisfying condition (C).

3.4 $W^\star$ Modules

If A is a $G^\star$ algebra satisfying condition (C), and if we have chosen connection elements, θ_A^a, then the homomorphism $\rho : W \to A$ makes A into an algebra over W; in particular, into a module over W. We want to generalize this notion slightly. To see why, consider the case of a compact Lie group, G acting freely on a non-compact manifold M. We can construct the connection forms $\theta_M^a \in \Omega(M)$ which satisfy (2.38). These forms will not vanish anywhere, in particular do not have compact support. But we may want to consider the algebra, $\Omega(M)_0$, of compactly supported forms on M. This algebra does not satisfy condition (C), but we can multiply any element of $\Omega(M)_0$ by any of the θ_M^a to get an element of $\Omega(M)_0$. In other words, $\Omega(M)_0$ is a module over W even though there is no $G^\star$ homomorphism of W into $\Omega(M)_0$. Armed with this motivation we make the following

Definition 3.4.1 *A $W^\star$ module is a $G^\star$ module B which is also a module over W in such a way that the map*

$$W \otimes B \to B, \qquad w \otimes b \mapsto wb$$

is a morphism of $G^\star$ modules. A $W^\star$ algebra is a $G^\star$ algebra which is a $W^\star$ module.

Recall that B_{hor} denotes the set of elements of B which satisfy (2.39). For each multi-index

$$I := (i_1, \ldots, i_r), \quad 1 \le i_1 < i_2 < \cdots < i_r \le n$$

let

$$\theta^I = \theta^{i_1} \cdots \theta^{i_r}$$

denote the corresponding monomial in the θ^a. Since each θ^a acts as an operator on B, the monomials θ^I act as operators on B.

Theorem 3.4.1 *Every element of a W^* module B can be written uniquely as a sum*

$$\theta^I h_I$$

with $h_I \in B_{\text{hor}}$.

Proof. We will prove the following lemma inductively:

Lemma 3.4.1 *Every element of B can be written uniquely as a sum*

$$\theta^J h_J$$

where

$$J = (j_1, \ldots, j_m), \quad 1 \le j_1 < \cdots < j_m \le k - 1$$

and

$$i_a h_J = 0, \quad a = 1, \ldots, k - 1.$$

The case $k = 1$ of the lemma says nothing and hence is automatically true. The case $k = n + 1$ is our theorem. So we assume that the lemma is true for $k - 1$ and prove it for k. Let

$$a_J := \iota_k h_J, \quad b_J := h_J - \theta^k a_J.$$

Then

$$\theta^J h_J = \theta^J b_J + \theta^J \theta^k a_J. \qquad \square$$

Identifying $\wedge(\theta^1, \ldots, \theta^n)$ with $\wedge(g^*) \subset W$, Theorem 3.4.1 says that the map

$$\wedge(g^*) \otimes B_{\text{hor}} \to B, \quad \theta^I \otimes h \mapsto \theta^I h \tag{3.20}$$

is bijective.

3.5   Bibliographical Notes for Chapter 3

1. Sections 3.1–3.3 are basically just an exposition of Weil's version of Chern-Weil theory. The first account of this theory to appear in print is contained in Cartan's paper: "Notions d'algèbre différentielle, ...".

2. One important G^* module to which this theory applies is the equivariant de Rham complex $\Omega_K(M)$, K being a (not necessarily compact) Lie group. If M is a $(G \times K)$-manifold on which G acts freely, there is a Chern-Weil map

$$S(g^*)^G \to H_K(M/G) \tag{3.21}$$

whose image is the ring of *equivariant* characteristic classes of M/G. (We will discuss (3.21) in more detail in Section 4.6.)

Chapter 4

The Weil Model and the Cartan Model

The results of the last chapter suggest that, for any G^* module B we take $B \otimes W$ as an algebraic model for the $X \times E$ of Chapter 1, and hence $H_{\text{bas}}(B \otimes W)$ as a definition of the equivariant cohomology of B. In fact, one of the purposes of this chapter will be to justify this definition. However the computation of $(B \otimes W)_{\text{bas}}$ is complicated. So we will begin with a theorem of Mathai and Quillen which shows how to find an automorphism of $B \otimes W$ which simplifies this computation. For technical reasons we will work with $W \otimes B$ instead of $B \otimes W$ and replace W by an arbitrary W^* module.

4.1   The Mathai-Quillen Isomorphism

Let A be W^* module and let B be a G^* module. Let

$$\theta^1, \ldots, \theta^n \in W_1, \quad \mu^1, \ldots, \mu^n \in W_2$$

be connection and curvature generators of the Weil algebra corresponding to a choice of basis, $\xi_1, \ldots, \xi_n$, of g. We define the degree zero endomorphism, $\gamma \in \text{End}(A \otimes B)$ by

$$\gamma := \theta^a \otimes \iota_a. \tag{4.1}$$

Notice that its definition is independent of the choice of basis. It is also invariant under the conjugation action of G.

   It is nilpotent; indeed

$$\gamma^{n+1} = 0$$

since every term in its expansion involves the application of $n + 1$ factors of θ. So $\phi \in \text{Aut}(A \otimes B)$ given by

$$\phi := \exp \gamma = 1 + \gamma + \frac{1}{2}\gamma^2 + \frac{1}{3!}\gamma^3 + \cdots \tag{4.2}$$

is a finite sum. The automorphism ϕ is known as the **Mathai-Quillen** isomorphism. It is an automorphism of G-modules.

For any $\beta \in \mathrm{End}(A \otimes B)$ we define

$$\mathrm{ad}\,\gamma(\beta) := [\gamma, \beta] = \gamma\beta - \beta\gamma$$

as usual. Notice that every term of

$$(\mathrm{ad}\,\gamma)^{2n+1}\beta$$

vanishes so $\mathrm{ad}\,\gamma$ is nilpotent and we have

$$\mathrm{Ad}(\phi)(\beta) := \phi\beta\phi^{-1} = \exp(\mathrm{ad}\,\gamma)\beta, \tag{4.3}$$

as this relation, $\exp \mathrm{ad} = \mathrm{Ad}\,\exp$, is true in any algebra of endomorphisms when the series on both sides converge.

We will now compute six instances of $(\mathrm{ad}\,\gamma)^k\beta$:

$$\mathrm{ad}\,\gamma(\iota_b \otimes 1) = -1 \otimes \iota_b \tag{4.4}$$

$$\mathrm{ad}\,\gamma(\nu \otimes \iota_b) = 0 \ \forall \nu \in A \tag{4.5}$$

$$(\mathrm{ad}\,\gamma)^2(\iota_b \otimes 1) = 0 \tag{4.6}$$

$$\mathrm{ad}\,\gamma(d) = -d\theta^a \otimes \iota_a + \theta^a \otimes L_a \tag{4.7}$$

$$(\mathrm{ad}\,\gamma)^2(d) = -c^k_{ab}\theta^a\theta^b \otimes \iota_k \tag{4.8}$$

$$(\mathrm{ad}\,\gamma)^3 d = 0. \tag{4.9}$$

Proof of (4.4). For $x \in A_k$, $y \in B_m$ we have

$$(\gamma \circ (\iota_b \otimes 1))\,(x \otimes y) = \gamma(\iota_b x \otimes y)$$
$$= (\theta^a \otimes \iota_a)(\iota_b x \otimes y)$$
$$= (-1)^{k-1}(\theta^a \iota_b x \otimes \iota_a y)$$

while

$$((\iota_b \otimes 1) \circ \gamma)\,(x \otimes y) = (-1)^k(\iota_b \otimes 1)\theta^a x \otimes \iota_a y$$
$$= (-1)^k x \otimes \iota_b y + (-1)^{k-1}\theta^a \iota_b x \otimes \iota_a y$$
$$= (1 \otimes \iota_b)(x \otimes y) + (-1)^{k-1}\theta^a \iota_b x \otimes \iota_a y.$$

Subtracting the second result from the first gives (4.4). $\square$

Of course, the role of the $x \otimes y$ in the above argument is just a crutch to remind us of the multiplication rule in

$$\mathrm{End}(A) \otimes \mathrm{End}(B) \subset \mathrm{End}(A \otimes B)$$

namely

$$(\alpha \otimes \beta)(\gamma \otimes \delta) = (-1)^{pq}\alpha\gamma \otimes \beta\delta \quad \text{if } \deg \beta = p, \ \deg \gamma = q$$

as we have checked by applying both sides to $x \otimes y$. This is just our usual rule: moving the β past the γ costs a sign; this time in the context of the tensor product of two algebras. So we can write the above argument more succinctly as

$$\begin{aligned}
\gamma(\iota_b \otimes 1) &= -\theta^a \iota_b \otimes \iota_a \\
(\iota_b \otimes 1)\gamma &= \iota_b \theta^a \otimes \iota_a \\
[\gamma, \iota_b \otimes 1] &= -[\theta^a, \iota_b] \otimes \iota_a \\
&= -\delta_b^a \otimes \iota_a \\
&= -1 \otimes \iota_b.
\end{aligned}$$

Proof of (4.5). Suppose that $\nu \in A_m$, and let $\pm = (-1)^m$. Then

$$\begin{aligned}
\gamma(\nu \otimes \iota_b) &= \pm \theta^a \nu \otimes \iota_a \iota_b \\
(\nu \otimes \iota_b)\gamma &= -\nu \theta^a \otimes \iota_b \iota_a \\
&= \mp \theta^a \nu \otimes \iota_b \iota_a
\end{aligned}$$

i.e.

$$[\gamma, \nu \otimes \iota_b] = \pm \theta^a \nu \otimes [\iota_a, \iota_b] = 0. \quad \square$$

Proof of (4.6). Equation (4.6) is an immediate consequence of (4.4) and (4.5).

Proof of (4.7). The d occurring in the left hand side of (4.7) is $d \otimes 1 + 1 \otimes d$. We have

$$[\theta^a \otimes \iota_a, d \otimes 1] = -[\theta^a, d] \otimes \iota_a = -d\theta^a \otimes \iota_a$$

while

$$[\theta^a \otimes \iota_a, 1 \otimes d] = \theta^a \otimes [\iota_a, d] = \theta^a \otimes L_a. \quad \square$$

Proof of (4.8). By (4.5) we have

$$\operatorname{ad}\gamma(d\theta^a \otimes \iota_a) = 0$$

so

$$(\operatorname{ad}\gamma)^2 d = (\operatorname{ad}\gamma)(\theta^a \otimes L_a)$$

by (4.7). We have

$$\begin{aligned}
[\gamma, \theta^a \otimes L_a] &= [\theta^b \otimes \iota_b, \theta^a \otimes L_a] \\
&= -\theta^b \theta^a \otimes \iota_b L_a - \theta^a \theta^b \otimes L_a \iota_b \\
&= -\theta^a \theta^b \otimes [L_a, \iota_b] \\
&= -c_{ab}^k \theta^a \theta^b \otimes \iota_k. \quad \square
\end{aligned}$$

Proof of (4.9). This follows from (4.5) and (4.8). □
Since γ is invariant under conjugation by G we know that

$$[\gamma, L_a \otimes 1 + 1 \otimes L_a] = 0.$$

It is an instructive exercise for the reader to prove this result by the above methods.

We now obtain the following theorem of Mathai-Quillen [MQ] and Kalkman [Ka].

Theorem 4.1.1 *The Mathai-Quillen isomorphism satisfies*

$$\phi\left(1 \otimes \iota_\xi + \iota_\xi \otimes 1\right)\phi^{-1} = \iota_\xi \otimes 1 \quad \forall \xi \in g \tag{4.10}$$

and

$$\phi d\phi^{-1} = d - \mu^k \otimes \iota_k + \theta^k \otimes L_k. \tag{4.11}$$

If A is a W^* algebra and B is a G^* algebra then ϕ is an algebra automorphism.

Proof of (4.10). Applying (4.3), (4.4) and (4.6) to $\xi = \xi_b$, the left-hand side of (4.10) is

$$
\begin{aligned}
(\exp \mathrm{ad}\,\gamma)\,(1 \otimes \iota_b + \iota_b \otimes 1) &= (1 \otimes \iota_b + \iota_b \otimes 1) + (\mathrm{ad}\,\gamma)(1 \otimes \iota_b + \iota_b \otimes 1) \\
&= (1 \otimes \iota_b + \iota_b \otimes 1) - 1 \otimes \iota_b \\
&= \iota_b \otimes 1. \ \square
\end{aligned}
$$

Proof of (4.11). Apply (4.3) and (4.7),(4.8) and (4.9). The left-hand side of (4.11) becomes

$$
\begin{aligned}
d + (\mathrm{ad}\,\gamma)d + \frac{1}{2}(\mathrm{ad}\,\gamma)^2 d &= d - d\theta^k \otimes \iota_k + \theta^k \otimes L_k - \frac{1}{2}c^k_{ab}\theta^a\theta^b \otimes \iota_k \\
&= d - \mu^k \otimes \iota_k + \theta^k \otimes L_k. \ \square
\end{aligned}
$$

The last statement in the theorem follows from the fact that γ is a derivation. This is true because in any algebra a derivation followed by multiplication by an element is again a derivation.

4.2   The Cartan Model

Equation (4.10) implies that ϕ carries $(A \otimes B)_{\mathrm{hor}}$, the horizontal subspace of $A \otimes B$, into $A_{\mathrm{hor}} \otimes B$:

$$\phi : (A \otimes B)_{\mathrm{hor}} \to A_{\mathrm{hor}} \otimes B. \tag{4.12}$$

Let us apply this to the case $A = W$. Then we may apply (3.11) which says that

$$W_{\mathrm{hor}} = \mathbf{C}[\mu^1, \ldots, \mu^n] \cong S(g^*)$$

and, according to (3.16), $d = d_W$ restricted to this subspace is

$$d = d_W = \theta^a L_a.$$

According to (4.11), ϕ conjugates $d = d_W \otimes 1 + 1 \otimes d_B$ into

$$\theta^a L_a \otimes 1 + 1 \otimes d_B + \theta^a \otimes L_a - \mu^a \otimes \iota_a = (\theta^a \otimes 1)(L_a \otimes 1 + 1 \otimes L_a) + 1 \otimes d_B - \mu^a \otimes \iota_a$$

on $W_{\text{hor}} \otimes B$. The space $(W \otimes B)_{\text{bas}}$ is just the space of invariant elements of $(W \otimes B)_{\text{hor}}$. Since ϕ is G-equivariant, it carries invariant elements into invariant elements and hence

$$\phi : (W \otimes B)_{\text{bas}} \to (S(g^*) \otimes B)^G. \tag{4.13}$$

The operator $L_a \otimes 1 + 1 \otimes L_a$ vanishes on invariant elements and hence

$$\phi d \phi^{-1} = 1 \otimes d_B - \mu^a \otimes \iota_a \tag{4.14}$$

on $(S(g^*) \otimes B)^G$.

For any G^* module B the space

$$C_G(B) := (S(g^*) \otimes B)^G \tag{4.15}$$

together with the differential

$$d_G : C_G(B) \to C_G(B), \quad d_G = 1 \otimes d_B - \mu^a \otimes \iota_a \tag{4.16}$$

is called the **Cartan model** for the equivariant cohomology of B. We can think of an element $\omega \in C_G(B)$ as being an equivariant polynomial map from g to B. With this interpretation, the element $(\mu^a \otimes \iota_a)\omega$ is the map

$$\xi \mapsto \iota_\xi \omega(\xi).$$

If ω is a homogeneous polynomial map then $\xi \mapsto \iota_\xi \omega(\xi)$ has polynomial degree one higher. So if

$$\omega \in S^k(g^*) \otimes A_\ell \quad \text{then} \quad \mu^a \otimes \iota_a \omega \in S^{k+1} \otimes A_{\ell-1}$$

and the total degree $2k + \ell$ is increased by one. From the point of view of polynomial maps the differential operator d_G is given by

$$d_G(\omega)(\xi) = d_B[\omega(\xi)] - \iota_\xi[\omega(\xi)] \tag{4.17}$$

and is of degree +1.

To summarize, we have proved the following fundamental theorem of Cartan:

Theorem 4.2.1 *The map ϕ carries $(W \otimes B)_{\text{bas}}$ into $C_G(B)$ and carries the restriction of $d = d_W \otimes 1 + 1 \otimes d_B$ into d_G. Thus*

$$H^* ((W \otimes B)_{\text{bas}}), d) = H^*(C_G(B), d_G). \tag{4.18}$$

The cohomology on the left is called the **Weil model** for the equivariant cohomology of B. We will justify its definition a little later on in this chapter by showing that $H^*((E \otimes B)_{\text{bas}}, d))$ is the same for any choice of acyclic W^* algebra, in particular $E = W$. So the thrust of Cartan's theorem is to say that the Cartan model gives the same cohomology as the Weil model.

4.3   Equivariant Cohomology of $W^\star$ Modules

In this section we assume that the group G is compact.

As we pointed out in Chapter 1, a key property of equivariant cohomology for topological spaces is the identity

$$H_G^*(M) = H^*(M/G)$$

if M is a topological space on which G acts freely.

The main result of this section is an algebraic analogue:

Theorem 4.3.1 *Let A be a $W^\star$ module and E an acyclic $W^\star$ algebra. Then*

$$H^* \left((A \otimes E)_{\text{bas}} \right) = H^*(A_{\text{bas}}). \tag{4.19}$$

We will prove this theorem using the Mathai-Quillen isomorphism, Theorem 4.1.1, taking $B = E$, an acyclic $W^\star$ algebra. So

$$\phi : (A \otimes E)_{\text{bas}} \to (A_{\text{hor}} \otimes E)^G$$

transforming the restriction of d to $(A \otimes E)_{\text{bas}}$ into

$$\delta = \delta_1 + \delta_2$$

where

$$\delta_1 := 1 \otimes d_E \tag{4.20}$$

and

$$\delta_2 := d_A \otimes 1 + \theta^a \otimes L_a - \mu^a \otimes \iota_a. \tag{4.21}$$

Define

$$C^i := \left(A_{\text{hor}} \otimes E^i \right)^G, \quad C^* := C^0 \oplus C^1 \oplus \cdots$$

so that

$$\delta_1 : C^i \to C^{i+1} \quad \text{and} \quad \delta_2 : C^i \to C^i \oplus C^{i-1},$$

and

$$\delta_1^2 = 0.$$

We can consider δ_1 as a differential on the complex C^*. We claim that

Lemma 4.3.1 *The cohomology groups of (C^*, δ_1) are given by*

$$H^0(C^*, \delta_1) = A_{\text{bas}}, \quad H^k(C^*, \delta_1) = 0, \quad k > 0. \tag{4.22}$$

Proof. Let

$$\tilde{C}^i := A_{\text{hor}} \otimes E^i, \quad \tilde{C}^* := \bigoplus \tilde{C}^i.$$

Since E is acyclic we have

$$H^0(\tilde{C}^*, 1 \otimes d_E) = A_{\text{hor}}, \quad H^k(\tilde{C}^*, 1 \otimes d_E) = 0, \quad \forall \, k > 0.$$

Since $C^i \subset \tilde{C}^i$, if $\tau \in C^k$, $k > 0$ satisfies $d\tau = 0$, we can find $\sigma \in \tilde{C}^{k-1}$ with $(1 \otimes d_E)\sigma = \tau$. Averaging σ over G, we may assume that it is G-invariant, i.e. lies in C^{k-1}. $\square$

Let

$$C_j := \bigoplus_{i \le j} C^i$$

so that

$$C_j \subset C_{j+1}, \quad (A_{\mathrm{hor}} \otimes E)^G = C^* = \bigcup C_j$$

gives an increasing filtration of $(A_{\mathrm{hor}} \otimes E)^G$ with $C_0 = A_{\mathrm{bas}}$.

To prove Theorem 4.3.1 it is enough to prove

Lemma 4.3.2 *If $\mu \in C_j$ satisfies*

$$\delta\mu = 0$$

then there is a $\nu \in C_{j-1}$ and an $a \in A_{\mathrm{bas}}$ with

$$\mu = \delta\nu + a \otimes 1. \tag{4.23}$$

Moreover a is unique up to a coboundary, i.e. if $\mu = 0$ in (4.23) then

$$a = d_A b, \quad b \in A_{\mathrm{bas}}.$$

Proof (by induction on j): Suppose $j = 0$. Then

$$\delta\mu = 0 \Rightarrow \delta_1\mu = 0 \Rightarrow \mu \in A_{\mathrm{bas}}$$

by Lemma 4.3.1. If $\nu \in C_0$ satisfies

$$\delta\nu + a \otimes 1 = 0$$

then $\delta_1\nu = 0$ so $\nu = -b \otimes 1$, $b \in A_{\mathrm{bas}}$ and

$$\delta_2\nu = (d_A \otimes 1)\nu + \theta^j \otimes L_j\nu - \mu^j \otimes \iota_j\nu = -(d_A b) \otimes 1$$

so

$$a = d_A b.$$

Now assume that $j > 0$ and we know the lemma for $j - 1$. Let $\mu \in C_j$ with $\delta\mu = 0$, and write $\mu = \mu_j + \cdots$ where $\mu_j \in C^j$ and the $\cdots$ lies in C_{j-1}. Then $\delta_1\mu_j = 0$ so $\mu = \delta_1\nu_{j-1}$ by Lemma 4.3.1. So

$$\mu = \delta\nu_{j-1} + w, \quad w = -\delta_2\nu + \cdots \in C_{j-1}.$$

Since $\delta\mu = 0$, we have $\delta w = 0$. We now may apply the inductive hypothesis to w. $\square$

The proof given above establishes an isomorphism between $H_{\mathrm{bas}}(A)$ and $H_G(A)$ in the case that A is a W^* module. The isomorphism might appear to depend on the actual structure of A as a W^* module, and not merely on its structure as a G^* module. However an analysis of the proof will show that this isomorphism depends only on the G^* structure. This will become even clearer in the next chapter when we examine the proof of Theorem 4.3.1 from the point of view of the Cartan model. Let

$$i : A_{\text{bas}} \to C_G(A) = (S(g^*) \otimes A)^G, \quad i : a \mapsto 1 \otimes a.$$

The map commutes with d and hence induces a homomorphism on cohomology. We will show that i, which depends only on the G^* structure, induces an isomorphism on cohomology by writing down a homotopy inverse for i. See Equations (5.9), (5.10), and (5.11) below.

4.4 $H\left((A \otimes E)_{\text{bas}}\right)$ does not depend on E

Let E and F be two acyclic W^* algebras. Then $A \otimes F$ is a W^* module and so

$$H^*_{\text{bas}}\left(A \otimes F \otimes E\right) = H^*_{\text{bas}}\left(A \otimes F\right)$$

by Theorem 4.3.1. Interchanging the role of E and F shows that

$$H^*_{\text{bas}}\left(A \otimes E\right) = H^*_{\text{bas}}\left(A \otimes F\right). \tag{4.24}$$

This provides the justification for using the Weil model

$$H_G(A) := H^*_{\text{bas}}\left(A \otimes W\right) \tag{4.25}$$

as the definition of equivariant cohomology, as we can replace W in this formula by any acyclic W^* algebra.

4.5 The Characteristic Homomorphism

Let $\phi : A \to B$ be a homomorphism of G^* algebras. We know that ϕ induces a homomorphism $\phi_* : H_G(A) \to H_G(B)$, and that the assignment $\phi \mapsto \phi_*$ is functorial. We also know that the equivariant cohomology of $\mathbf{C}$, regarded as a trivial G^*-module is given by $H_G(\mathbf{C}) = H_{\text{bas}}(W) = S(g^*)^G$. Suppose that A is a (unital) G^* algebra, so has a unit element $1 = 1_A$ which is G-invariant (and hence basic). The map

$$\kappa : \mathbf{C} \to A, \quad z \mapsto z1_A$$

is a homomorphism of G^* modules, and hence induces a homomorphism

$$\kappa_G : S(g^*)^G \to H_G(A).$$

This map is called the **characteristic homomorphism** or the **Chern-Weil map**. The elements of the image of κ_G are known as the **characteristic**

classes of $H_G(A)$. In the case that $A = \Omega(M)$ where M is a manifold, it has the following alternative description: The unique map

$$M \to \text{pt.}$$

of M onto the unique, connected, zero-dimensional manifold, pt., induces, by functoriality, a map

$$H_G^*(\text{pt.}) \to H_G^*(M).$$

Hence, if G acts freely on M, a map

$$H_G^*(\text{pt.}) \to H^*(X), \quad X := M/G.$$

Since

$$H_G^*(\text{pt.}) = H_G(\Omega(\text{pt.})) = H_G(\mathbf{C}) = S(g^*)^G,$$

this identifies our map κ_G as a map

$$S(g^*)^G \to H^*(X).$$

This is the usual Chern-Weil map. We will discuss the structure of $S(g^*)^G$ for various important groups G in Chapter 8. This will then yield a description of the more familiar characteristic classes. To compute κ_G in the Weil model, observe that the map

$$\kappa \otimes \text{id} : S(g^*)^G = \mathbf{C} \otimes S(g^*)^G \to (A \otimes S(g^*))_{\text{bas}}$$

given by tensoring by 1_A maps $S(g^*)^G$ into closed elements in the Weil model, and passing to the cohomology gives κ_G.

Every element of the image of $\kappa \otimes \text{id}$ is fixed by the Mathai-Quillen homomorphism, ϕ, and so, in the Cartan model, $\kappa \otimes \text{id}$ is again the map given by tensoring the invariant elements of $S(g^*)$ by 1_A. Passing to the corresponding cohomology classes then gives κ_G.

4.6 Commuting Actions

Let M and K be Lie groups. Suppose that $G = M \times K$ as a group, with the corresponding decomposition $g = m \oplus k$ as Lie algebras. Then $\tilde{m}$ and $\tilde{k}$ can be regarded as subalgebras of $\tilde{g}$ with $m_{-1} \oplus m_0$ commuting with $k_{-1} \oplus k_0$. Also, we have the natural decomposition of Weil algebras,

$$W(g) = W(m) \otimes W(k), \quad d_{W(g)} = d_{W(m)} \otimes 1 + 1 \otimes d_{W(k)}.$$

Any G^* module A can be thought of as an M^* module and as a K^* module. The space of elements of A which are basic for the M^* action, call it $A_{\text{bas}_{M^*}}$, is a submodule for the K^* action and vice versa. We have

$$A_{\text{bas}_{G^*}} = (A_{\text{bas}_{K^*}})_{\text{bas}_{M^*}}.$$

in the obvious notation. We can apply this to $A \otimes W(g) = A \otimes W(m) \otimes W(k)$. Suppose that A, and hence $A \otimes W(m)$ is a a $W(k)^*$ module. Then, by Theorem 4.3.1,

$$
\begin{aligned}
H_G(A) &:= H\left([(A \otimes W(m) \otimes W(k))_{\mathrm{bas}_{M^*}}]_{\mathrm{bas}_{K^*}}, d\right) \\
&= H\left([A_{\mathrm{bas}_{K^*}} \otimes W(m)]_{\mathrm{bas}_{M^*}}, d\right) \\
&=: H_M(A_{\mathrm{bas}_{K^*}}).
\end{aligned}
$$

We conclude that

$$
H_G(A) = H_M(A_{\mathrm{bas}_{K^*}}). \tag{4.26}
$$

If A is also a $W(m)^*$ module (when considered as an M^* module), we conclude that

$$
H_K(A_{\mathrm{bas}_{M^*}}) = H_M(A_{\mathrm{bas}_{K^*}}). \tag{4.27}
$$

In the case that M is compact we can describe (4.26) in terms of the Cartan map. Indeed, suppose that the θ^i are the connection forms that make A into a $W(k)^*$ module for the K^* action. Since M and K commute, we may average these θ's over M using the M action to obtain ones that are M-invariant. Then

$$
C_G(A) = \left[S(k^*) \otimes [S(m^*) \otimes A]^M\right]^K
$$

and

$$
d_{C(G)} = d_{C(K)}(d_{C(M)}) = 1 \otimes d_{C(M)} - \nu^j \iota(\eta_j),
$$

where $\{\eta_1, \ldots \eta_r\}$ is a basis of k and $\{\nu^1, \ldots, \nu^r\}$ the corresponding dual basis of k^*, and where $d_{C(K)}$ is the Cartan d relative to K^* of $d_{C(M)}$, the Cartan d relative to M^* of A. This cohomology is isomorphic to cohomology relative to $d_{C(M)}$ of

$$
[S(m^*) \otimes A_{\mathrm{bas}_{K^*}}]^M
$$

which is just $H_M(A_{\mathrm{bas}_{K^*}})$.

In particular, we have the characteristic homomorphism

$$
\kappa_K : S(k^*)^K \to H_M(A_{\mathrm{bas}_{K^*}}). \tag{4.28}
$$

The image of κ_K is called the ring of **M-equivariant characteristic classes**.

4.7 The Equivariant Cohomology of Homogeneous Spaces

Let K be a closed subgroup of the compact group G and apply (4.27) with $G \times K$ playing the role of G, where G acts on itself from the left and K from the right, giving commuting free actions of G and K on G. We conclude that

$$
H_G(G/K) = H_K(G/G) = S(k^*)^K \tag{4.29}
$$

computing the equivariant cohomology of a homogeneous space.

4.8   Exact Sequences

Let G be a compact Lie group and

$$\cdots \to A_{i-1} \to A_i \to A_{i+1} \to \cdots \tag{4.30}$$

be an exact sequence of G^* modules. Tensoring with $S(g^*)$ gives an exact sequence

$$\cdots \to A_{i-1} \otimes S(g^*) \to A_i \otimes S(g^*) \to A_{i+1} \otimes S(g^*) \to \cdots$$

and hence a sequence

$$\cdots \to (A_{i-1} \otimes S(g^*))^G \to (A_i \otimes S(g^*))^G \to (A_{i+1} \otimes S(g^*))^G \to \cdots.$$

We claim that this sequence is also exact. Indeed, suppose that ν is in the kernel of

$$(A_i \otimes S(g^*))^G \to (A_{i+1} \otimes S(g^*))^G.$$

Then there is a

$$\mu \in A_{i-1} \otimes S(g^*)$$

whose image is ν. Since ν is G invariant, the image of $a\mu$ is also ν for any $a \in G$. Hence, averaging all the $a\mu$ with respect to Haar measure gives an element of $(A_{i-1} \otimes S(g^*))^G$ whose image is ν.

We have thus proved

Theorem 4.8.1 *An exact sequence (4.30) of G^* modules gives rise to an exact sequence of Cartan complexes*

$$\cdots \to C_G(A_{i-1}) \to C_G(A_i) \to C_G(A_{i+1}) \to \cdots \tag{4.31}$$

In particular, consider a short exact sequence

$$0 \to A \to B \to C \to 0$$

of G^* modules. By Theorem 4.8.1 we get a short exact sequence

$$0 \to C_G(A) \to C_G(B) \to C_G(C) \to 0$$

of complexes and hence a long exact sequence in cohomology

$$\cdots \to H_G^k(A) \to H_G^k(B) \to H_G^k(C) \to H_G^{k+1}(A) \to \cdots. \tag{4.32}$$

4.9   Bibliographical Notes for Chapter 4

1. Most of the material in this chapter is due to Cartan and is contained in Sections 5-6 of "La transgression dans un groupe de Lie...".   A

word of warning: These two sections (which consist of five brief paragraphs) don't make for easy reading: they contain the definition of the Weil model (page 62, lines 20-23), the definition of the Cartan model (page 63, lines 30-33), a proof of the equivalence of these two models (page 63, lines 19-37), the definition of what we're really calling a "$W^\star$ module" (page 62, line 32), a proof of the isomorphism,

$$H\left((A \otimes E)_{\text{bas}}\right) = H(A_{\text{bas}})$$

(page 63, lines 7-17) and most of the results which we'll discuss in the next chapter (page 64, lines 1-21).

2. The Mathai-Quillen isomorphism is implicitly in Cartan, is much more explicitly described in section 5 of [MQ] and is made even more explicit in Kalkman's thesis [Ka]. Our version of Mathai-Quillen is a somewhat simplified form of that in [Ka].

3. There are some very interesting variants of the Cartan model, due to Berline and Vergne and their co-authors: An element, ρ, of the Cartan complex $(\Omega(M) \otimes S(g^*))^G$, can be regarded as an equivariant mapping

$$\rho : g \to \Omega(M) \tag{4.33}$$

which depends polynomially on g, and its equivariant coboundary, $d_G \rho$, can be defined to be the mapping

$$(d_G \rho)(\xi) = d(\rho(\xi)) - \iota_\xi \rho(\xi). \tag{4.34}$$

This definition, however, doesn't require ρ to be a *polynomial* function of ξ. One can for instance define the equivariant cohomology of M with C^∞ coefficients:

$$H_G^\infty(M)$$

to be the cohomology of the complex of *smooth* mappings, (4.33), with the coboundary operator (4.34) (c.f. [BV], [DV], [BGV]) and one can define an equivariant cohomology of M with *distributional* coefficients

$$H_G^{-\infty}(M)$$

by allowing the mappings (4.33) to be *distributional* functions of g. (See [KV].)

4. The proof of Theorem 4.3.1 can be streamlined a bit by using the spectral sequence techniques that we describe in Chapter 6. By (4.20) and (4.21) $(A \otimes E)_{\text{bas}}$ is a bicomplex with coboundary operators δ_1, and δ_2, and by lemma 4.3.1 its spectral sequence collapses at the E_2 stage with $E_2^{0,q} = H^q(A_{\text{bas}})$ and $E_2^{p,q} = 0$ for $p \neq 0$.

Chapter 5

Cartan's Formula

In this chapter we do some more detailed computations in the Cartan model. Recall that

$$C_G(A) = (S(g^*) \otimes A)^G.$$

It will be convenient, in order not to have to carry too many tensor product signs, to identify $S(g^*) \otimes A$ with the space of A-valued polynomials. If $\xi_1, \ldots, \xi_n$ is a basis of g, we will let $x^1, \ldots x^n$ denote the corresponding coordinates, i.e. the corresponding dual basis. (So we are temporarily using x^i instead of the μ^i used in $W(g)$ for pedagogical reasons.) The Cartan differential in this notation is given by

$$(d_G \alpha)(x) := d[\alpha(x)] - x^r \iota_r [\alpha(x)].$$

We set

$$\partial_r := \frac{\partial}{\partial x^r}.$$

In the current notation, the polynomial

$$f(x) = \sum_I a_I x^I$$

(where $I = (i_1, \ldots, i_n)$ is a multi-index) is identified with the element

$$\sum_I x^I \otimes a_I = \sum \mu^I \otimes a_I \in S(g^*) \otimes A \subset W(g) \otimes A.$$

The fact that A is a $W(g)$-module means that we have an evaluation map

$$W(g) \otimes A \to A$$

sending

$$x^I \otimes a_I \mapsto \mu^I a_I,$$

so we will denote the image of $f(x)$ under this map by $f(\mu)$. In other words,

$$f(\mu) := \sum \mu^I a_I.$$

This can also be written as follows: Let

$$S := \mu^a \partial_a. \tag{5.1}$$

So S is an operator on A-valued polynomials. Let

$$\rho : S(g^*) \otimes A \to A$$

be evaluation of a polynomial at 0, so

$$\rho(f) := f(0) \in A.$$

Then

$$f(\mu) = \rho\left[(\exp S)f\right]. \tag{5.2}$$

Put "geometrically," the operator $\exp S$ is just the "translation" $x \to x + \mu$ in f and ρ has the effect of setting $x = 0$. In other words, we are taking the Taylor expansion of f at 0 with μ "plugged in".

The **basic subcomplex** $C^{0,0} \subset C_G(A)$ is defined to consist of those maps which satisfy

$$\iota_r \omega = 0, \quad \partial_s \omega = 0, \quad \forall r, s.$$

The second equations say that $\omega \in C_G(A)$ is a constant map, and so is a G-invariant element of A, while the first equations say that ω is horizontal. So $C^{0,0}(A) = A_{\text{bas}}$ when G is connected.

5.1   The Cartan Model for $W^\star$ Modules

Let us give an alternative proof, using the Cartan model, of Theorem 4.3.1, namely that for a $W^\star$ module, A, we have the formula

$$H_G^*(A) = H_{\text{bas}}^*(A). \tag{5.3}$$

Suppose that A is a $W^\star$ module so that there are connection elements θ^r and their corresponding curvature elements μ^r acting on A, corresponding to a choice of basis of g.

Define the operators $\mathcal{K}, E, R$ on $C_G(A)$ by

$$
\begin{aligned}
\mathcal{K} &:= -\theta^r \partial_r \\
E &:= x^r \partial_r + \theta^r \iota_r \\
R &:= d\theta^r \partial_r.
\end{aligned}
$$

We want to think of E as the supersymmetric version of the Euler operator, where the $\{\theta^r\}$ are thought of as odd variables. In our current notational attempt to eliminate tensor product signs, we write the $1 \otimes d_A$ occurring on the right hand side of (4.16) simply as d and the $\mu^a \otimes \iota_a$ as $x^a \iota_a$ so that (4.16) becomes

$$d_G = d - x^a \iota_a.$$

For any $\alpha \in C_G(A)$ we have

$$
\begin{aligned}
(d\mathcal{K} + \mathcal{K}d)\alpha &= -d\theta^r \partial_r \alpha + \theta^r \partial_r d\alpha - \theta^r \partial_r d\alpha \\
&= -R\alpha \\
(-x^r \iota_r)\mathcal{K}\alpha &= (x^r \iota_r)\theta^s \partial_s \alpha \\
&= x^r \partial_r \alpha - x^r \theta^s \iota_r \partial_r \alpha \quad \text{since } \iota_r \theta^s = \delta_r^s \\
\mathcal{K}(-x^r \iota_r)\alpha &= (-\theta^s \partial_s)(-x^r \iota_r)\alpha \\
&= \theta^r \iota_r \alpha + x^r \theta^s \iota_r \partial_s \alpha
\end{aligned}
$$

so adding all terms shows that

$$d_G \mathcal{K} + \mathcal{K} d_G = E - R. \tag{5.4}$$

It follows that

$$d_G(E - R) = (E - R)d_G. \tag{5.5}$$

The operators $x^r \partial_r$ and $\theta^r \iota_r$ commute and map $C_G(A)$ into itself, so we have the simultaneous eigenspace decomposition:

$$C_G(A) = \bigoplus C^{p,q},$$

where p is the degree as a polynomial in x and q represents the "vertical degree" in the sense that an element of $C^{*,q}(A)$ is a sum of terms of the form

$$\theta^{i_1} \cdots \theta^{i_q} \cdot \omega \quad \text{where } \omega \in S(g^*) \otimes A_{\text{hor}}.$$

In other words, $C^{p,q}$ is the image of $\wedge^q(g^*) \otimes S^p(g^*) \otimes A_{\text{hor}}$ under the evaluation map $W(g) \otimes A \to A$.

In particular, this notation is consistent with the previous notation in that $C^{0,0}$ consists of basic elements of A. Also, introduce the filtration corresponding to polynomials of degree at most p:

$$C^{(p)} := \bigoplus_{\ell \leq p} C^{\ell,*}.$$

We have

- $\mathcal{K}$ lowers filtration degree by 1,

- d_G raises filtration degree by 1,

- E preserves filtration degree,

- R lowers filtration degree, (and is nilpotent)

- $E = (p+q)\mathrm{id}$ on $C^{p,q}$.

So $E - R$ is invertible on $\bigoplus_{(p,q)\neq(0,0)} C^{p,q}$.
Let
$$\pi : C_G(A) \to C^{(0,0)}$$
denote projection along $\bigoplus_{(p,q)\neq(0,0)} C^{p,q}$ so
$$\pi = 0 \text{ on } C^{p,q}, \quad (p,q) \neq (0,0), \quad \pi = \mathrm{id} \text{ on } C^{0,0}.$$

Notice that on $\bigoplus C^{p,0} = (S(g^*) \otimes A_{\mathrm{hor}})^G$ the operator π is just evaluation of a polynomial at 0, in other words, it coincides with the operator ρ defined in the introduction to this chapter. In particular, formula (5.2) holds on $(S(g^*) \otimes A_{\mathrm{hor}})^G$ with $\rho = \pi$.

The operator
$$J := E + \pi - R$$
is invertible, and we can write (5.4) as
$$d_G \mathcal{K} + \mathcal{K} d_G = J - \pi. \tag{5.6}$$

Let
$$F := (E + \pi)^{-1}$$
so
$$J = (I - RF)(E + \pi)$$
and define
$$U := J^{-1} = F(I - RF)^{-1} = F\left(I + RF + (RF)^2 + \cdots\right). \tag{5.7}$$

The series on the right is locally finite since RF lowers filtration degree. Let
$$Q := \mathcal{K}U. \tag{5.8}$$

We will prove that
$$d_G Q + Q d_G = I - \pi U. \tag{5.9}$$

If we let $i : C^{0,0} \to C_G(A)$ denote the inclusion, then $i \circ \pi = \pi$ so we can write (5.9) as
$$d_G Q + Q d_G = I - i \circ (\pi U). \tag{5.10}$$

On the other hand $U = \mathrm{id}$ on $C^{0,0}$ so
$$I = (\pi U) \circ i \tag{5.11}$$

on $C^{0,0} = A_{\mathrm{bas}}$. Thus the maps i and πU are homotopy inverses of one another, and hence induce isomorphisms on cohomology. In other words, the formula (5.9) implies that
$$H(A_{\mathrm{bas}}, d) = H(C_G(A), d_G) = H_G(A),$$

and that this isomorphism depends only on the G^* structure and not on the W^* structure.

Proof of (5.9). It follows from (5.5) that

$$d_G J - J d_G = d_G \pi - \pi d_G = [d_G, \pi]$$

so multiplying on the right and left by $U = J^{-1}$ we get

$$U d_G - d_G U = U[d_G, \pi]U.$$

Now d_G maps $C^{0,0}$ into itself and so $d_G \circ \pi : C_G(A) \to C^{0,0}$ and so does π and hence $\pi \circ d_G$. So $[d_G, \pi] : C_G \to C^{0,0}$ and $U = \text{id}$ on $C^{0,0}$. So we can simplify the last equation to

$$[U, d_G] = [d_G, \pi]U.$$

Since $\mathcal{K} = 0$ on $C^{0,0}$ we have

$$K[d_G, \pi] = 0.$$

Then

$$
\begin{aligned}
d_G Q + Q d_G &= d_G \mathcal{K} U + \mathcal{K} U d_G \\
&= d_G \mathcal{K} U + \mathcal{K} d_G U + \mathcal{K}[d_G, \pi]U \\
&= (d_G \mathcal{K} + \mathcal{K} d_G)U \\
&= (J - \pi)U \\
&= I - \pi U
\end{aligned}
$$

proving (5.9). □

5.2   Cartan's Formula

We now do a more careful analysis which will lead to a rather explicit formula for the operator πU. We have

$$\pi U = \pi + \pi R F + \pi (R F)^2 + \cdots$$

from (5.7) since F preserves the bigradation and equals the identity on $C^{0,0}$. We may write

$$R = S - T$$

where

$$S = \mu^a \partial_a : \ C^{i,j} \to C^{i-1,j}.$$

and

$$T = \frac{1}{2} c_{ij}^a \theta^i \theta^j \partial_a : \ C^{i,j} \to C^{i-1,j+2}.$$

Now T increases the q degree by two and S does not decrease it so we can write

$$\pi U = \pi + \pi SF + \pi(SF)^2 + \cdots.$$

The operator S decreases total degree, $\ell = p + q$ by one and F takes the value $1/\ell$ on elements of total degree $\ell \neq 0$. Also $\pi = 0$ on elements of total degree not equal to 0. Hence

$$\pi(SF)^\ell = \frac{1}{\ell!}\pi S^\ell$$

and so

$$\pi U = \pi \exp S.$$

S, and hence $\exp S$ does not change the q degree. So

$$\pi U = \pi \exp S \circ \mathrm{Hor}$$

where Hor denotes the projection $C_G(A) \to \bigoplus_p C^{p,0}$. Now every element of $C^{p,0}$ is a sum of terms of the form $x^I \omega$ where x^I is a monomial of total degree $|I| = p$ and ω a horizontal element of A. On such a term we have $\pi S^\ell = 0$ unless $\ell = p$ and then $\frac{1}{\ell!}\pi S^\ell(x^I \omega) = \mu^I \omega$. So

$$\pi \exp S(x^I \omega) = \mu^I \omega,$$

as we observed in the introduction, equation (5.2).

In any event, we have proved the following theorem due to Cartan:

Theorem 5.2.1 *The Cartan operator πU is the composition of the projection operator*

$$\mathrm{Hor} : C_G(A) \to \bigoplus_i C^{i,0} = (S(g^*) \otimes A_{\mathrm{hor}})^G$$

and the map

$$(S(g^*) \otimes A_{\mathrm{hor}})^G \to A_{\mathrm{bas}}$$

coming from the W^* module structure, i.e. the "evaluation map"

$$x^I \otimes a \to \mu^I a.$$

Let us see what the Cartan map looks like in the case of manifolds. Let M be a G-manifold on which G acts freely, let $X := M/G$, and let $\pi : M \to X$ be the map which assigns to every $m \in M$ its G-orbit. We can think of this situation as a principal G-bundle. Equipping this bundle with a connection, we get curvature forms

$$\mu^i \in \Omega(M)_{\mathrm{hor}}$$

and a G-equivariant map

$$S(g^*)^G \otimes \Omega(M) \to \Omega(M)_{\mathrm{hor}}, \qquad x^I \otimes \omega \mapsto \mu^I \wedge \omega_{\mathrm{hor}}.$$

From π we get a bijection

$$\pi^* : \Omega(X) \to \Omega(M)_{\text{bas}}.$$

Combining the first map, restricted to the invariants, with the inverse of the second map, we obtain a map

$$(S(g^*) \otimes \Omega(M))^G \to \Omega(X).$$

This is the manifold version of πU in the theorem. The restriction of this map to $S(g^*)^G$ is the Chern-Weil map at the level of forms. It induces a map on cohomology

$$S(g^*)^G \to H^*(X)$$

which is the Chern-Weil map that we discussed in section 4.5. For

$$p = p(x^1, \ldots, x^n) \in S(g^*)^G$$

its image under κ_G is given by the cohomology class of

$$p(\mu^1, \ldots, \mu^n).$$

This map is well defined and independent of the choice of connection.

5.3   Bibliographical Notes for Chapter 5

1. A close textual reading of "La transgression dans une groupe de Lie...", page 64, lines 1-21 reveals that the proof of the Cartan isomorphism: $H_G(A) = H(A_{\text{bas}})$, which we've given in this chapter is that envisaged by Cartan. We are grateful to Hans Duistermaat for explaining this proof to us. The construction of the homotopy operator in Section 5.1 is based on some unpublished notes of his as is the beautiful identity, $\pi U = \pi \exp S$ in section 5.2.

2. Not only can the Cartan formula be used to reprove the isomorphism: $H_G(A) = H(A_{\text{bas}})$ but it has many other applications besides. We will discuss a couple of these in Chapter 8 and Chapter 10. In Chapter 8 we will use the Cartan formula to give simple proofs of two well-known theorems in symplectic geometry: the Duistermaat-Heckmann Theorem and the minimal coupling theorem. In Chapter 10 we will use it to obtain a formula of Mathai-Quillen for the Thom form of an equivariant vector bundle in terms of the curvature forms of the bundle.

Chapter 6

Spectral Sequences

We begin this chapter with a review of the theory of spectral sequences in the special context of double complexes. We then apply these results to equivariant cohomology: We will show that if a G^* morphism between two G^* modules induces an isomorphism on cohomology it induces an isomorphism on equivariant cohomology. Given a G^* module A, we will discuss the structure of $H_G(A)$ as an $S(g^*)^G$-module, and show that if the spectral sequence associated with A collapses at its E_1 stage then $H_G(A)$ is free as an $S(g^*)^G$-module. Finally, we will prove an abelianization theorem which says that

$$H_G(A) = H_T(A)^{\mathbf{W}}$$

where T is a Cartan subgroup (maximal torus) of G and $\mathbf{W}$ its Weyl group.

6.1 Spectral Sequences of Double Complexes

A **double complex** is a bigraded vector space

$$C = \bigoplus_{p,q \in \mathbf{Z}} C^{p,q}$$

with coboundary operators

$$d : C^{p,q} \to C^{p,q+1}, \quad \delta : C^{p,q} \to C^{p+1,q}$$

satisfying

$$d^2 = 0, \quad d\delta + \delta d = 0, \quad \delta^2 = 0.$$

The associated **total complex** is defined by

$$C^n := \bigoplus_{p+q=n} C^{p,q}$$

with coboundary $d + \delta : C^n \to C^{n+1}$.

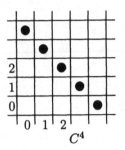

$$C^4$$

Let

$$C_k^n := \bigoplus_{p+q=n,\ p\geq k} C^{p,q},$$

so C_k^n consists of all elements of C^n which live to the right of a vertical line.

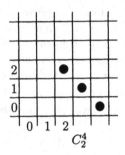

$$C_2^4$$

Let

$$Z_k^n := \{z \in C_k^n,\ (d+\delta)z = 0\}, \qquad B^n := (d+\delta)C^{n-1}.$$

The map

$$Z_k^n \to Z_k^n/(B^n \cap Z_k^n)$$

gives an decreasing filtration

$$\cdots \supset H_{k-1}^n \supset H_k^n \supset H_{k+1}^n \supset \cdots$$

of the cohomology group $H^n(C, d+\delta)$. We denote the successive quotients by $H^{k,n-k}$ and let

$$\operatorname{gr} H^n := \bigoplus_k H^{k,n-k} \qquad H^{k,n-k} := H_k^n/H_{k+1}^n. \tag{6.1}$$

The spectral sequence we are about to describe is a scheme for computing these quotients starting from the cohomology groups of the "vertical complexes"

$$(C^{r,*}, d).$$

We will construct a sequence of complexes (E_r, δ_r) such that each E_{r+1} is the cohomology of the preceding one, $E_{r+1} = H(E_r, \delta_r)$, and (under suitable hypotheses) the "limit" of these complexes are the quotients $H^{p,q}$.

To get started on all this, we will need a better description of these quotients: Any element of C^n has all its elements on the (anti-)diagonal line $\bigoplus_{i+j=n} C^{i,j}$ and will have a "leading term" at position (p, q) where p denotes the smallest i where the component does not vanish. Let $Z^{p,q}$ denote the set of such components of cocycles at position (p, q). In other words, $Z^{p,q}$ denotes the set of all $a \in C^{p,q}$ with the property that the system of equations

$$
\begin{aligned}
da &= 0 \\
\delta a &= -da_1 \\
\delta a_1 &= -da_2 \\
\delta a_2 &= -da_3 \\
&\vdots
\end{aligned}
\tag{6.2}
$$

admits a solution

$$(a_1, a_2, \ldots) \quad \text{where} \quad a_i \in C^{p+i, q-i}.$$

In other words, $a \in Z^{p,q} \subset C^{p,q}$ can be ratcheted by a sequence of zigzags to any position below a on the (anti-)diagonal line ℓ through a, where $\ell = \{(i, j) \mid i + j = p + q\}$:

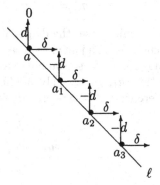

In the examples we will be encountering,

$$C^{i,j} = 0 \quad \text{for } i + j = p + q, \ |i - j| > m_\ell \tag{6.3}$$

for some m_ℓ and hence the system (6.2) will be solvable for all i provided that it is solvable for a bounded range of i. To repeat: the equations (6.2) say that the element

$$z := a \oplus a_1 \oplus a_2 \oplus \cdots$$

lies in Z_p^n and has "leading coefficient" a.

Let $B^{p,q} \subset C^{p,q}$ consist of all b with the property that the system of equations

$$
\begin{aligned}
dc_0 + \delta c_{-1} &= b \\
dc_{-1} + \delta c_{-2} &= 0 \\
dc_{-2} + \delta c_{-3} &= 0 \\
dc_{-3} + \delta c_{-4} &= 0 \\
&\ \ \vdots
\end{aligned}
\tag{6.4}
$$

admits a solution

$$(c_0, c_{-1}, c_{-2}, \ldots) \quad \text{with } c_{-i} \in C^{p-i,q+i-1}.$$

Once again, if the boundedness condition (6.3) holds, it suffices to solve these equations for a bounded range of i.

It is easy to check that the quotients $H^{p,q}$ defined in (6.1) can also be described as

$$H^{p,q} = Z^{p,q}/B^{p,q}. \tag{6.5}$$

Let us try to compute these quotients by solving the system (6.2) inductively.

Let $Z_r^{p,q} \subset C^{p,q}$ consist of those $a \in C^{p,q}$ for which the first $r-1$ of the equations (6.2) can be solved. In other words, $a \in Z_r^{p,q}$ if and only if it can be joined by a sequence of zigzags to an element $a_{r-1} \in C^{p_r,q_r}$ where

$$(p_r, q_r) = (p + r - 1, q - r + 1) \tag{6.6}$$

is the point $r - 1$ units down the diagonal ℓ from (p, q).

When can such an a be joined by a sequence of r zigzags to an element of C^{p_r+1,q_r-1}? In order to do so, we may have to retrace our steps and replace the partial solution $(a_1, \ldots, a_{r-1})$ by a different partial solution $(a'_1, \ldots, a'_{r-1})$ so that the differences, $a''_i = a'_i - a_i$ satisfy

$$
\begin{aligned}
da''_1 &= 0 \\
\delta a''_1 + da''_2 &= 0 \\
&\ \ \vdots
\end{aligned}
\tag{6.7}
$$

$$\vdots$$

$$\delta a''_{r-2} + da''_{r-1} = 0$$

and to zigzag one step further down we need an $a''_r \in C^{p_r+1,q_r-1}$ such that

$$\delta a_{r-1} = -\delta a''_{r-1} - da''_r.$$

Let us set

$$b := \delta a_{r-1}, \quad c_0 := -a''_r, \quad c_{-i} := -a''_{r-i}, \quad i = 1, \ldots, r-1$$

and $c_{-i} = 0$ for $i \geq r$. So let us define

$$B_r^{p,q} \subset B^{p,q}$$

be the set of all $b \in C^{p,q}$ for which there is a solution of (6.4) with $c_{-i} = 0$ for $i \geq r$. Then we have proved

Theorem 6.1.1 *Let $a \in Z_r^{p,q}$. Then*

$$a \in Z_{r+1}^{p,q} \quad \Leftrightarrow \quad \delta a_{r-1} \in B_r^{p_r+1,q_r}$$

for any solutions $(a_1, \ldots, a_{r-1})$ of the first $r - 1$ equations of (6.2).

Notice that since δa_{r-1} satisfies the system of equations

$$
\begin{aligned}
\delta a_{r-1} &=: \quad b \\
\delta a_{r-2} + d a_{r-1} &= \quad 0 \\
&\vdots \\
&\vdots \\
da &= \quad 0
\end{aligned}
\tag{6.8}
$$

we see that δa_{r-1} itself is in $B_{r+1}^{p_r+1,q_r}$. Indeed, *every* element of $B_{r+1}^{p_r+1,q_r}$ can be written as a sum of the form

$$\delta a_{r-1} + dc \tag{6.9}$$

with $c \in C^{p_r+1,q_r-1}$ and $(a, a_1, \ldots, a_{r-1})$ a solution of the first $r-1$ equations of (6.2). From this we can draw a number of conclusions:

1. Let

$$E_r^{p,q} := Z_r^{p,q}/B_r^{p,q}. \tag{6.10}$$

   Since

$$\delta a_{r-1} \in B_{r+1}^{p_r+1,q_r} \subset Z^{p_r+1,q_r} \subset Z_r^{p_r+1,q_r},$$

   we see that δa_{r-1} projects onto an element

$$\delta_r a \in E_r^{p_r+1,q_r}. \tag{6.11}$$

   Theorem 6.1.1 can be rephrased as saying that an $a \in Z_r^{p,q}$ lies in $Z_{r+1}^{p,q}$ if and only if $\delta_r a = 0$.

2. It is clear that $\delta_r a$ only depends on the class of a modulo $B^{p,q}$. Since $B^{p,q} \supset B_r^{p,q}$ we can consider δ_r to be a map

$$\delta_r : E_r^{p,q} \rightarrow E_r^{p+r,q-r+1}. \tag{6.12}$$

3. By (6.9), the image of this map is the projection of $B_{r+1}^{p+r,q-r+1}$ into $E_r^{p+r,q-r+1}$ and, by Theorem 6.1.1, the kernel of this map is the projection of $Z_{r+1}^{p,q}$ into $E_r^{p,q}$. Thus the sequence

$$\cdots \xrightarrow{\delta_r} E_r^{p,q} \xrightarrow{\delta_r} \cdots$$

has the property that $\ker \delta_r \supset \operatorname{im} \delta_r$ and

$$(\ker \delta_r)/(\operatorname{im} \delta_r) = E_{r+1}^{p,q} \qquad (6.13)$$

in position (p, q).

In other words, the sequence of complexes

$$(E_r, \delta_r), \quad r = 1, 2, 3, \ldots$$

has the property that

$$H(E_r, \delta_r) = E_{r+1}. \qquad (6.14)$$

By construction, these complexes are bigraded and δ_r is of bidegree

$$(r, -(r-1)).$$

Moreover, if condition (6.3) is satisfied for all diagonals, ℓ, the "spectral sequence" eventually stabilizes with

$$E_r^{p,q} = E_{r+1}^{p,q} = \cdots$$

for r large enough (depending on p and q). Moreover, this "limiting" complex, according to (6.5), is given by

$$E_\infty^{p,q} = \lim E_r^{p,q} = H^{p,q}. \qquad (6.15)$$

6.2   The First Term

The case $r = 1$ of (6.14) is easy to describe: By definition,

$$E_1^{p,q} = H^q(C^{p,*}, d) \qquad (6.16)$$

is the vertical cohomology of each column. Moreover, since d and δ commute, one gets from δ an induced map on cohomology

$$H^q(C^{p,*}, d) \to H^q(C^{p+1,*}, d)$$

and this is the induced map δ_1. So we have described (E_1, δ_1) and hence E_2. The δ_r for $r \geq 2$ are more complicated but can be thought of as being generalizations of the connecting homomorphisms in the long exact sequence in cohomology associated with a short exact sequence of cochain complexes as is demonstrated by the following example:

6.3   The Long Exact Sequence

Let (C, d, δ) be a double complex with only three non-vanishing columns corresponding to $p = 0, 1$, or 2, In other words, we assume that

$$C^{p,q} = 0, \quad p \neq 0, 1, 2.$$

The E_1 term of this spectral sequence is

$$0 \to H\left(C^{0,*}, d\right) \xrightarrow{i} H\left(C^{1,*}, d\right) \xrightarrow{j} H\left(C^{2,*}, d\right) \to 0$$

with $i = \delta_1$, $j = \delta_1$. Thus

$$E_2^{0,q} = \ker i : H^q\left(C^{0,*}, d\right) \to H^q\left(C^{1,*}, d\right)$$

$$E_2^{1,q} = \frac{\ker j : H^q\left(C^{1,*}, d\right) \to H^q\left(C^{2,*}, d\right)}{\operatorname{im} i : H^q\left(C^{0,*}, d\right) \to H^q\left(C^{1,*}, d\right)}$$

$$E_2^{2,q} = \frac{H^q\left(C^{2,*}, d\right)}{\operatorname{im} j : H^q\left(C^{1,*}, d\right) \to H^q\left(C^{2,*}, d\right)}$$

The coboundary operator, δ_2 maps $E_2^{0,q} \to E_2^{2,q-1}$ and vanishes on $E_2^{1,q}$ and on $E_2^{2,q}$. For $r > 2$ we have $\delta_r = 0$, so this spectral sequence "collapses at its E_3 stage".

Suppose now that the rows of the double complex are exact. In other words, suppose that

$$0 \to C^{0,*} \xrightarrow{\delta} C^{1,*} \xrightarrow{\delta} C^{2,*} \to 0$$

is a short exact sequence of complexes. If we interchange rows and columns, we get a double complex whose *columns* are exact, hence a double complex having its $E_1 = 0$. Thus we must have $E_3 = 0$ in our original double complex. Hence we must have

$$E_3^{p,q} = 0$$

and

$$\delta_2 : E_2^{0,q+1} \to E_2^{2,q}$$

is an isomorphism. So if we define the "connecting homomorphism" κ as

$$\kappa := \delta_2^{-1} : E_2^{2,q} \to E_2^{0,q+1}$$

we get the long exact sequence

$$\cdots \xrightarrow{\kappa} H^q(C^{0,*}, d) \xrightarrow{i} H^q(C^{1,*}, d) \xrightarrow{j} H^q(C^{2,*}, d) \xrightarrow{\kappa} H^{q+1}(C^{0,*}, d) \xrightarrow{i} \cdots$$

in cohomology.

6.4   Useful Facts for Doing Computations

6.4.1   Functorial Behavior

Let (C, d, δ) and (C', d', δ') be double complexes, and $\rho : C \to C'$ a morphism of double complexes of bidegree (m, n) which intertwines d with d', and δ with δ'. This give rise to a cochain map

$$\rho : (C, d + \delta) \to (C', d' + \delta')$$

of degree $m + n$. It induces a map $\rho_{\natural}$ on the total cohomology:

$$\rho_{\natural} : H(C, d + \delta) \to H(C', d' + \delta')$$

of degree $m + n$ and consistent with the filtrations on both sides. Similarly ρ maps the cochain complex $(C^{p,*}, d)$ into the cochain complex $((C')^{p+m,*}, d')$ and hence induces a map on cohomology

$$\rho_1 : E_1 \to E_1'$$

of bidegree (m, n) which intertwines δ_1 with δ_1'. Inductively we get maps

$$\rho_r : (E_r, \delta) \to (E_r', \delta_r'). \tag{6.17}$$

Here ρ_{r+1} is the map on cohomology induced from ρ_r where, we recall, $E_{r+1} = H(E_r, \delta_r)$. It also is clear that

Theorem 6.4.1 *If the two spectral sequences converge, then*

$$\lim \rho_r = \operatorname{gr} \rho_{\natural}. \tag{6.18}$$

In particular,

Theorem 6.4.2 *If ρ_r is an isomorphism for some $r = r_0$ then it is an isomorphism for all $r > r_0$ and so, if both spectral sequences converge, then $\rho_{\natural}$ is an isomorphism.*

6.4.2   Gaps

Sometimes a pattern of zeros among the $E_r^{p,q}$ allows for easy conclusions. Here is a typical example:

Theorem 6.4.3 *Suppose that $E_r^{p,q} = 0$ when $p + q$ is odd. Then the spectral sequence "collapses at the E_r stage", i.e. $E_r = E_{r+1} = \cdots$.*

Proof. $\delta_r : E_r^{p,q} \to E_r^{p+r,q-r+1}$ so changes the parity of $p + q$. Thus either its domain or its range is 0. So $\delta_r = 0$. $\square$

6.4.3   Switching Rows and Columns

We have already used this technique in our discussion of the long exact sequence. The point is that switching p and q, and hence d and δ does not change the total complex, but the spectral sequence of the switched double complex can be quite different from that of the original. We will use this technique below in studying the spectral sequence that computes equivariant cohomology. Another illustration is Weil's famous proof of the de Rham theorem, [We]:

Theorem 6.4.4 *Let (C, d, δ) be a double complex all of whose columns are exact except the $p = 0$ column, and all of whose rows are exact except the $q = 0$ row. Then*

$$H(C^{0,*}, d) = H(C^{*,0}, \delta). \tag{6.19}$$

Proof. The E_1 term of the spectral sequence associated with (C, d, δ) has only one non-zero column, the column $p = 0$, and in that column the entries are the cohomology groups of $(C^{0,*}, d)$. Hence $\delta_r = 0$ for $r \geq 2$ and

$$H(C, d + \delta) = H(C^{0,*}, d).$$

Switching rows and columns yields

$$H(C, d + \delta) = H(C^{*,0}, \delta).$$

Putting these two facts together produces the isomorphism stated in the theorem.   □

6.5   The Cartan Model as a Double Complex

Let G be a compact Lie group and let $A = \bigoplus A^k$ be a G^* module. Its Cartan complex

$$C_G(A) = (S(g^*) \otimes A)^G$$

can be thought of as a double complex with bigrading

$$C^{p,q} := \left(S^p(g^*) \otimes A^{q-p}\right)^G, \tag{6.20}$$

and with vertical and horizontal operators given by

$$d := 1 \otimes d_A \tag{6.21}$$

and

$$\delta := -\mu^a \otimes \iota_a. \tag{6.22}$$

Notice that in the bigrading (6.20) the subspace $(S^p(g^*) \otimes A^m)^G$ has bidegree $(p, m + p)$ and hence total degree $2p + m$ which is the grading that we have been using on $(S(g^*) \otimes A)^G$ as a commutative superalgebra.

If $A^k = 0$ for $k < 0$, which will be the case in all our examples, then the double complex satisfies our diagonal boundedness condition (6.3), and so, under this assumption, the spectral sequence associated to the double complex (6.20),(6.21), and (6.22) will converge.

We begin by evaluating the E_1 term:

Theorem 6.5.1 *The E_1 term in the spectral sequence of (6.20) is*

$$(S(g^*) \otimes H(A))^G.$$

More explicitly

$$E_1^{p,q} = \left(S^p(g^*) \otimes H^{q-p}(A)\right)^G. \tag{6.23}$$

Proof. The complex $C = (S(g^*) \otimes A)^G$ with boundary operator $1 \otimes d_A$ sits inside the complex

$$(S(g^*) \otimes A, 1 \otimes d_A) \tag{6.24}$$

and (by averaging over the group) the cohomology groups of $(C, 1 \otimes d_A)$ are just the G-invariant components of the cohomology of (6.24) which are the appropriately graded components of $S(g^*) \otimes H(A)$. $\square$

To compute the right-hand side of (6.23) we use (cf. Remark 3 in Section 2.3.1)

Proposition 6.5.1 *The connected component of the identity in G acts trivially on $H(A)$.*

Proof. It suffices to show that all the operators L_a act trivially on on $H(A)$. But

$$L_a = \iota_a d + d\iota_a$$

says that L_a is chain homotopic to 0 in A. $\square$

So we get

Theorem 6.5.2 *If G is connected then*

$$E_1^{p,q} = S^p(g^*)^G \otimes H^{q-p}(A). \tag{6.25}$$

We now may apply the gap method to conclude

Theorem 6.5.3 *If G is connected and $H^p(A) = 0$ for p odd, then the spectral sequence of the Cartan complex of A collapses at the E_1 stage.*

Proof. By (6.25), $E_1^{p,q} = 0$ when $p + q$ is odd. $\square$

Let M be a G-manifold on which G acts freely and let A be the de Rham complex, $A = \Omega(M)$ so that $H_G(A) = H^*(X)$ where $X = M/G$. Theorem 6.5.1 gives a spectral sequence whose E_1 term is

$$S(g^*)^G \otimes H^*(M)$$

and whose E_∞ terms is a graded version of $H^*(X)$. The topological version of this spectral sequence is the Leray-Serre spectral sequence associated with the fibration (1.4). See [BT], page 169. The notation there is a bit different. They use a slightly different bicomplex so that their E_2 term corresponds to our E_1 term.

6.6 $H_G(A)$ as an $S(g^*)^G$-Module

We assume from now on that G is compact and connected so that (6.25) holds. If $f \in S^m(g^*)^G$, the multiplication operator

$$u \otimes a \mapsto fu \otimes a$$

is a morphism of the double complex (C, d, δ) given by (6.20), (6.21), and (6.22) of bidegree (m, m) and so it induces a map of $H_G(A)$ into itself. In other words, we have given $H_G(A)$ the structure of an $S(g^*)^G$-module. Also, all the E_r's in the spectral sequence become $S(g^*)^G$-modules. Under the identification (6.25) of E_1 this module structure is just multiplication on the left factor of the right hand side of (6.25), which shows that E_1 is a free $S(g^*)^G$-module. Now $S(g^*)^G$ is Noetherian; see [Chev]. So if $H(A)$ is finite dimensional, all of its subquotients, in particular all the E_r's are finitely generated as $S(g^*)^G$-modules. Since the spectral sequence converges to a graded version of $H_G(A)$, we conclude

Theorem 6.6.1 *If dim $H(A)$ is finite, then $H_G(A)$ is finitely generated as an $S(g^*)^G$-module.*

Another useful fact that we can extract from this argument is:

Theorem 6.6.2 *If the spectral sequence of the Cartan double complex collapses at the E_1 stage, then $H_G(A)$ is a free $S(g^*)^G$-module.*

Proof. Equation (6.25) shows that E_1 is free as an $S(g^*)^G$-module, and if the spectral sequence collapses at the E_1 stage, then $E_1 \cong$ gr $H_G(A)$ and this isomorphism is an isomorphism of $S(g^*)^G$-modules by Theorem 6.4.2. So gr $H_G(A)$ is a free $S(g^*)^G$-module and hence is so is $H_G(A)$. $\square$

6.7   Morphisms of $G^\star$ Modules

Let

$$\rho : A \to B$$

be a morphism of degree zero between two G^* modules. We get an induced morphism between the corresponding Cartan double complexes and hence induced maps

$$\rho_* : H(A, d) \to H(B, d)$$

on the ordinary cohomology and

$$\rho_\sharp : H_G(A) \to H_G(B)$$

on the equivariant cohomology. From Theorems 6.4.2 and 6.5.2 we conclude:

Theorem 6.7.1 *If the induced map ρ_* on ordinary cohomology is bijective, then so is the induced map $\rho_\sharp$ on equivariant cohomology.*

6.8   Restricting the Group

Suppose that G is a compact connected Lie group and that K is a closed subgroup of G (not necessarily connected). We then get an injection of Lie algebras

$$k \to g$$

and of superalgebras

$$\tilde{k} \to \tilde{g}$$

so every G^* module becomes a K^* module by restriction. Also, the injection $k \to g$ induces a projection

$$g^* \to k^*$$

which extends to a map

$$S(g^*) \to S(k^*)$$

and then to a map

$$(S(g^*) \otimes A)^G \to (S(k^*) \otimes A)^K$$

which is easily checked to be a morphism of complexes, in fact of double complexes

$$C_G(A) \to C_K(A).$$

We thus get an restriction mapping

$$H_G(A) \to H_K(A)$$

and also a restriction morphism at each stage of the corresponding spectral sequences. Now since G acts trivially on $H(A)$, being connected, and K is a subgroup of G, we conclude that K also acts trivially on $H(A)$ even though it need not be connected. In particular the conclusion of Theorem 6.5.2 applies to K as well, and hence the restriction morphism on the E_1 level is just the restriction applied to the left hand factors in

$$S(g^*)^G \otimes H(A) \to S(k^*)^K \otimes H(A).$$

Therefore, by Theorem 6.4.2 we conclude:

Theorem 6.8.1 *Suppose that the restriction map*

$$S(g^*)^G \to S(k^*)^K$$

is bijective. Then the restriction map

$$H_G(A) \to H_K(A)$$

in equivariant cohomology is bijective.

Unfortunately, there is only one non-trivial example we know of for which the hypothesis of the theorem is fulfilled, but this is a very important example. Let T be a Cartan subgroup of G and let $K = N(T)$ be its normalizer. The quotient group

$$\mathbf{W} = K/T$$

is the Weyl group of G. It is a finite group so the Lie algebra of K is the same as the Lie algebra of T. Since T is abelian, its action on t^*, hence on $S(t^*)$, is trivial. So

$$S(k^*)^K = S(t^*)^K = S(t^*)^{\mathbf{W}}.$$

According to a theorem of Chevalley, see for example [Helg] (Chapter X Theorem 6.1), the restriction

$$S(g^*)^G \to S(t^*)^{\mathbf{W}}$$

is bijective so Theorem 6.8.1 applies.

We can do a bit more: From the inclusion $T \to K$ we get a morphism of double complexes

$$C_K(A) \to C_T(A)^{\mathbf{W}}$$

which induces a morphism

$$H_K(A) \to H_T(A)^{\mathbf{W}}$$

and also a morphism at each stage of the spectral sequences. At the E_1 level this is just the identity morphism

$$S(t^*)^{\mathbf{W}} \otimes H(A) \to S(t^*)^{\mathbf{W}} \otimes H(A)$$

and hence another application of Theorem 6.4.2 yields

$$H_K(A) = H_T(A)^{\mathbf{W}}.$$

Putting this together with the isomorphism coming from Theorem 6.8.1 we obtain the important result:

Theorem 6.8.2 *Let G be a connected compact Lie group, T a maximal torus and $\mathbf{W}$ its Weyl group. Then for any G^*-module A we have*

$$H_G(A) \cong H_T(A)^{\mathbf{W}}. \qquad (6.26)$$

This result can actually be strengthened a bit: The tensor product

$$C_K(A) \otimes_{S(t^*)^{\mathbf{W}}} S(t^*)$$

is also a bicomplex since the coboundary operators on $C_K(A)$ are $S(t^*)^{\mathbf{W}}$-module morphisms. Moreover there is a canonical morphism

$$C_K(A) \otimes_{S(t^*)^{\mathbf{W}}} S(t^*) \to C_T(A), \qquad a \otimes f \mapsto fa. \qquad (6.27)$$

The spectral sequence associated with the bicomplex $C_K(A) \otimes_{S(t^*)^{\mathbf{W}}} S(t^*)$ converges to

$$H_K(A) \otimes_{S(t^*)^{\mathbf{W}}} S(t^*).$$

From (6.27) one gets a morphism of spectral sequences which is an isomorphism at the E_1 level. Hence, at the E_∞ level we have

$$H_G(A) \otimes_{S(t^*)^{\mathbf{W}}} S(t^*) \cong H_T(A). \tag{6.28}$$

Theorem 6.8.2 has an interesting application in topology: the **splitting principle**. Let M be a G-manifold on which G acts freely. Let

$$X := M/G, \qquad Y := M/T$$

and let

$$\pi : Y \to X$$

be the map which assigns to every T-orbit the corresponding G-orbit. This is a differentiable fibration with typical fiber G/T. One gets from it a map

$$\pi^* : H^*(X) \to H^*(Y).$$

Moreover there is a natural action of the Weyl group $\mathbf{W}$ on Y which leaves fixed the fibers of π. Hence π^* maps $H^*(X)$ into $H^*(Y)^{\mathbf{W}}$.

Theorem 6.8.3 *The map*

$$\pi^* : H^*(X) \to H^*(Y)^{\mathbf{W}}$$

is a bijection.

Proof. This follows from (6.26) and the identifications

$$\begin{aligned} H_G(M) &\cong H^*(X) \\ H_T(M) &\cong H^*(Y) \end{aligned}$$

given by the Cartan isomorphism. $\square$

We can sharpen this result. The theorem of Chevalley cited above also asserts that $S(t^*)^{\mathbf{W}}$ is finitely generated, and is, in fact, a polynomial ring in finitely many generators. Let $x^1, \ldots, x^r$ be a basis of t^* and let

$$p_i(x^1, \ldots, x^r), \qquad i = 1, \ldots k$$

be generators of $S(t^*)^{\mathbf{W}}$. The Chern-Weil map

$$\kappa_T : S(t^*) \to H^*(Y)$$

maps $x^1, \ldots, x^r$ into cohomology classes $\sigma^1, \ldots, \sigma^r$ and the Chern-Weil map

$$\kappa_G : S(g^*) = S(t^*)^{\mathbf{W}} \to H^*(X)$$

maps $p_1, \ldots, p_k$ into cohomology classes $c_1, \ldots c_k$ satisfying

$$\pi^* c_i - p_i(\sigma^1, \ldots, \sigma^r) = 0. \tag{6.29}$$

From (6.28) one easily deduces

Theorem 6.8.4 $H^*(Y)$ *is the quotient of the ring*

$$H^*(X)[\sigma^1, \ldots, \sigma^r]$$

by the ideal generated by the expressions on the left-hand side of (6.29).

For more details on the splitting principle in topology, see Section 8.6 below.

6.9 Bibliographical Notes for Chapter 6

1. There are several other versions of the theory of spectral sequences besides the "bicomplex version" that we've presented here. For the *Massey* version (the spectral sequence associated with an exact couple) see [Ma] or [BT], and for the *Koszul* version (the spectral sequence associated with a filtered cochain complex) see [Go] or [Sp]. The oldest and most venerable topological example of a spectral sequence is the Serre-Leray spectral sequence associated with a fibration

$$F \hookrightarrow X \to B.$$

   If B is simply connected, the E_2 term of this spectral sequence is the tensor product of $H(B)$ and $H(F)$ and the E_∞ term is a graded version of $H(X)$. (For a description of this sequence as the spectral sequence of a bicomplex, see [BT] page 169.)

2. For example, the E_2 term of the spectral sequence associated with the fibration (1.6) is

$$H(E/G) \otimes H(M), \tag{6.30}$$

   and its E_∞ term is $H_G(M)$. Since G acts freely on E, $H(E/G) = H_G(E)$; and since E is contractible. $H_G(E) = H_G(\text{pt,}) = S(g^*)^G$; so (6.30) is equal to:

$$S(g^*)^G \otimes H(M)$$

   i.e., is equal to the E_1 term of the spectral sequence associated with the Cartan model

$$(S(g^*) \otimes \Omega(M))^G .$$

   (The E_∞ term is, of course, the same: $H_G(M)$.)

3. If this spectral sequence collapses at its E_1 stage M is said to be *equivariantly formal*. Goresky, Kottwitz, and MacPherson examine this property in detail (in a much broader context than ours) in [GKM] and derive a number inequivalent sufficient conditions for it to hold. In particular they prove

Theorem 6.9.1 *Suppose the ordinary homology of M, $H_k(M, \mathbf{R})$, is generated by classes which are representable by cycles, ξ, each of which is invariant under the action of K. Then M is equivariantly formal.*

We will discuss some of their other necessary and sufficient conditions in the Bibliographical Notes to chapter 11.

4. An important example of equivariant formality was discovered by Kirwan [Ki] and, independently, by Ginzburg [Gi]: M is equivariantly formal if it is compact and admits an equivariant symplectic form.

5. If M is equivariantly formal, then as an $S(g^*)^G$ module

$$H_G(M) = H(M) \otimes S(g^*)^G$$

by the remarks in section 6.6. Tensoring this identity with the trivial $S(g^*)^G$ module $\mathbf{C}$ gives

$$H(M) = H_G(M) \otimes_{S(g^*)^G} \mathbf{C}$$

expressing the ordinary de Rham cohomology of M in terms of its equivariant cohomology.

6. Let G be a compact Lie group, T a Cartan subgroup and K a closed subgroup of G with $T \subset K \subset G$. Combining (6.28) with (4.29) we get

$$H_T(G/K) = S(t^*)^{\mathbf{W}_K} \otimes_{S(t^*)^{\mathbf{W}_G}} S(t^*)$$

where $\mathbf{W}_K$ and $\mathbf{W}_G$ are the Weyl groups of K and G. Using note 5, we get the following expression for the ordinary de Rham cohomology of G/K:
$$H(G/K) = S(t^*)^{\mathbf{W}_K} / m S(t^*)^{\mathbf{W}_K}$$

where $m = (S(t^*)^{\mathbf{W}_G})_+$, the maximal ideal of $S(t^*)^{\mathbf{W}_G}$ at zero, cf. [GHV] Chapter X Theorem XI (p. 442).

7. The relation of the splitting principle described in Section 6.8 to the usual splitting principle for vector bundles will be explained in Section 8.6.

8. In [AB] section 2, Atiyah and Bott give a purely topological proof of Theorem 6.8.3 and then, by reversing the sense of our argument, deduce Theorem 6.8.1.

Chapter 7

Fermionic Integration

Fermionic integration was introduced by Berezin [Be] and is part of the standard repertoire of elementary particle physicists. It is not all that familiar to mathematicians. However it was used by Mathai and Quillen [MQ] in their path breaking paper constructing a "universal Thom form". In this chapter we will develop enough of Berezin's formalism to reproduce the Mathai-Quillen result. We will also discuss the Fermionic Fourier transform and combine Bosonic and Fermionic Fourier transforms into a single "super" Fourier transform. We will see that there is an equivariant analogue of compactly supported cohomology which can be obtained from the Koszul complex by using this super Fourier transform, and use this to explain the Mathai-Quillen formula. In Chapter 10 we will apply these results to obtain localization theorems in equivariant cohomology.

7.1 Definition and Elementary Properties

Let V be an d-dimensional real vector space equipped with an oriented volume element, that is, a chosen basis element, vol, of $\wedge^d V$. A **preferred basis** $\{\psi^1, \ldots, \psi^d\}$ of V is a basis such that

$$\psi^1 \wedge \ldots \wedge \psi^d = \text{vol}.$$

Elements $f \in \wedge V$ are thought of as "functions in the odd variables ψ" in that every such element can be written as

$$f = f(\psi^1, \ldots, \psi^d) = f_I \psi^I. \tag{7.1}$$

When $I = (1, \ldots, n)$,

$$\psi^I = \text{vol},$$

and the coefficient f_{vol} is called the **Berezin integral** or the **Fermionic integral** of f:

$$\int f(\psi^1, \ldots, \psi^d) d\psi := f_{vol} \quad \text{when } f = f_I \psi^I. \tag{7.2}$$

Various familiar formulas in the integral calculus have their analogues for this notion of integration, but with certain characteristic changes:

7.1.1   Integration by Parts

If $\tau \in V^*$ then interior product by τ is a derivation of degree -1 of $\wedge V$:

$$\iota_\tau : \wedge^k V \to \wedge^{k-1} V.$$

In particular, $\iota_\tau f$ has no component in degree n, and hence

$$\int [\iota_\tau f](\psi^1, \ldots, \psi^d) d\psi = 0.$$

If $\tau_1, \ldots, \tau_d$ is the basis dual to $\psi^1, \ldots, \psi^d$, let us denote the operation ι_{τ_i} by $\partial/\partial\psi^i$:

$$\frac{\partial}{\partial\psi^i} := \iota_{\tau_i}.$$

So we have

$$\int \frac{\partial}{\partial\psi^i} f(\psi^1, \ldots, \psi^d) d\psi = 0. \tag{7.3}$$

We can apply this to a product: Recall that

$$(\wedge V)_0 = \bigoplus \wedge^{2k} V, \quad (\wedge V)_1 = \bigoplus \wedge^{2k+1} V$$

is the $\mathbf{Z}/2\mathbf{Z}$-gradation of $\wedge V$. Then using (7.3) and the fact that $\partial/\partial\psi^i$ is a derivation, we get

$$\int \frac{\partial u}{\partial\psi^i} v \, d\psi = -(-1)^r \int u \frac{\partial v}{\partial\psi^i} d\psi \quad \text{if } \deg u = r, \tag{7.4}$$

where $\deg u = 0$ if $u \in (\wedge V)_0$ and $\deg u = 1$ if $u \in (\wedge V)_1$.

7.1.2   Change of Variables

Consider a "linear change of variables" of the form

$$\psi^i \mapsto \sum a^i_j \psi^j.$$

This then induces, by multiplication, a map

$$\wedge V \to \wedge V$$

and, in particular

$$\text{vol} \mapsto (\det A)\, \text{vol} \quad \text{where } A = \left(a_j^i\right).$$

More abstractly, we are considering a linear map of $V \to V$ and extending to an algebra homomorphism of $\wedge V \to \wedge V$ which exists and is uniquely determined by the universal property of the exterior algebra. In any event, we can write the preceding equality as

$$\int f(\textstyle\sum a_j^1 \psi^j, \ldots, \sum a_j^d \psi^j)d\psi = \det A \int f(\psi^1, \ldots, \psi^d)d\psi. \tag{7.5}$$

Notice an important differences between this Fermionic "change of variables" law and the standard ("Bosonic") rule. In ordinary (Bosonic) integration the rule for a linear change of variables would have a $(\det A)^{-1}$ on the right-hand side. We will have occasion to use both Bosonic and Fermionic linear changes of variables in what follows.

7.1.3   Gaussian Integrals

Let $d = 2m$ and let $q \in \wedge^2 V$ so

$$q = q(\psi^1, \ldots, \psi^d) = q_{ij}\psi^i\psi^j, \quad q_{ij} = -q_{ji}.$$

in terms of a preferred basis. Then, writing

$$Q := (q_{ij})$$

we have

$$\int \exp -\frac{1}{2}q(\psi^1, \ldots, \psi^d)d\psi = (\det Q)^{\frac{1}{2}}. \tag{7.6}$$

Here exp is the exponential in the exterior algebra, given by its usual power series formula which becomes a polynomial since q is nilpotent.

Proof. We may apply a linear transformation with determinant one to bring Q to the normal form

$$Q = \begin{pmatrix} \lambda_1 J & 0 & \cdots & 0 \\ 0 & \lambda_2 J & \cdots & 0 \\ \vdots & \vdots & \ddots & \vdots \\ 0 & 0 & \cdots & \lambda_m J \end{pmatrix} \quad \text{where} \quad J := \begin{pmatrix} 0 & -1 \\ 1 & 0 \end{pmatrix}, \tag{7.7}$$

so that

$$-\frac{q}{2} = \lambda_1 \psi^1 \psi^2 + \lambda_2 \psi^3 \psi^4 + \cdots + \lambda_m \psi^{2m-1} \psi^{2m}.$$

Then the component of $\exp -\frac{q}{2}$ lying in $\wedge^d(V) = \wedge^{2m}(V)$ is (by the multinomial formula)

$$(-2)^m \frac{1}{m!} q^m = \frac{1}{m!} \cdot m! \lambda_1 \cdots \lambda_m \psi^1 \cdots \psi^{2m} = \lambda_1 \cdots \lambda_m \, \text{vol} \,.$$

So

$$\int \exp -\frac{q}{2}(\psi^1, \ldots, \psi^d) d\psi = \lambda_1 \cdots \lambda_m.$$

On the other hand, the determinant of Q as given in the above normal form is clearly

$$\lambda_1^2 \cdots \lambda_d^2. \quad \square$$

Notice the contrast with the "Bosonic" Gaussian integral,

$$\int_V e^{-q(x,x)/2} dx, \qquad q(x,y) := q_{ij} x^i y^j$$

where $q_{ij} = q_{ji}$ is symmetric and must be positive-definite for the integral to converge. The result is

$$(2\pi)^{-m} |\det Q|^{-\frac{1}{2}}.$$

The factors of 2π are conventional — due to our choice of normalization of Lebesgue measure so that the unit cube have volume one. The key difference between the Bosonic and Fermionic Gaussian integrals is that $|\det Q|^{\frac{1}{2}}$ occurs in the denominator in the Bosonic case and in the numerator in the Fermionic case.

If V carries a scalar product we can think of $q \in \wedge^2 V$ as an element of $g = o(n)$, and the formula 7.6 becomes the supersymmetric definition of the *Pfaffian* of q. (For an alternative non-supersymmetric definition see §8.2.3 below.) Thus we can write (7.6) as

$$\int \exp -\frac{q}{2}(\psi^1, \ldots, \psi^d) d\psi = \text{Pfaff } q. \tag{7.8}$$

7.1.4 Iterated Integrals

Let A be an arbitrary commutative superalgebra and consider elements of $A \otimes \wedge V$ as the exterior algebra analogue of functions with values in A. So we may consider the expression (7.1) as an element of $A \otimes \wedge V$ where now the "coefficients" f_I are elements of A. We then can use exactly the same definition, (7.2), for the integral; the end result of the integration yielding an element of A. The operator ι_τ is interpreted as $1 \otimes \iota_\tau$ and then the integration-by-parts formula (7.4) continues to hold, where $\deg u$ now means the $\mathbf{Z}/2\mathbf{Z}$-degree of u as an element of $A \otimes \wedge V$. The operation of integration is even or odd depending on the dimension of the vector space V.

In particular, we can take $A = \wedge U$ where U is a second vector space with preferred volume element. We have

$$\wedge(U \oplus V) = \wedge U \otimes \wedge V$$

and let us choose the volume element on $U \oplus V$ so that

$$\text{vol}_{U \oplus V} = \text{vol}_U \cdot \text{vol}_V$$

in the obvious notation.

7.1.5 The Fourier Transform

In this subsection we wish to develop the Fermionic analogue of the Fourier transform, using Fermionic integration. We begin by recalling some basic facts about the classical (Bosonic) Fourier transform: Let V be a d-dimensional real vector space space with volume element du and suppose that we have chosen linear coordinates $u^1, \ldots u^d$ so that

$$du = du^1 \cdots du^d.$$

We let $\mathcal{S}(V)$ denote the Schwartz space of rapidly decreasing smooth functions on V.

1. For $f \in \mathcal{S}(V)$, its Fourier transform $\hat{f} \in \mathcal{S}(V^*)$ is defined by

$$\hat{f}(y) := \left(\frac{1}{2\pi} \right)^{d/2} \int_V f(u) e^{-i\langle u, y \rangle} du$$

where $\langle \ , \ \rangle$ denotes the pairing between V and V^*.

2. If $y_1, \ldots, y_d$ denote the coordinates on V^* dual to $u^1, \ldots, u^d$, then integration by parts gives

$$\left(\widehat{\frac{\partial f}{\partial u^j}} \right)(y) = iy_j \hat{f}(y) \tag{7.9}$$

and

3. Differentiation under the integral sign gives

$$(\widehat{u^j f}) = i \frac{\partial \hat{f}}{\partial y_j}. \tag{7.10}$$

4. The Fourier inversion formula asserts that

$$f(u) = \left(\frac{1}{2\pi} \right)^{d/2} \int_{V^*} \hat{f}(y) e^{i\langle u, y \rangle} dy$$

and hence, if the map

$$f \mapsto \hat{f}$$

is denoted by F, that

5.

$$F^2(f)(u) = f(-u).$$

There are various choices of convention that have been made here — the use of $-i$ in 1) and hence of i in 4) and the placement of the factors of 2π.

We now wish to develop an analogue in the Fermionic case. We do this by taking the A of the preceding subsection to be $\wedge(V^*)$, the exterior algebra of the dual space of V. Here A is generated by the basis $\tau_1, \ldots, \tau_d$ of V^*, dual to the basis $\psi^1, \ldots, \psi^d$ of V. Define the (Fermionic) **Fourier transform**

$$F : \wedge V \to \wedge V^*, \quad f \mapsto \hat{f}$$

by

$$Ff(\tau_1, \ldots, \tau_d) = \hat{f}(\tau_1, \ldots, \tau_d) = \int \exp(i\tau_k \psi^k) f(\psi^1, \ldots, \psi^d) d\psi. \quad (7.11)$$

The map F is clearly linear. Define $\omega \in \wedge(V^*) \otimes \wedge(V)$ by

$$\omega := i\tau_k \psi^k.$$

The definition of ω is independent of the the choice of basis and we have defined the Fourier transform as

$$F(f) = \int f \exp(\omega) d\psi.$$

Notice that ω is an even element of $\wedge(V^*) \otimes \wedge(V)$ and that

$$\frac{\partial}{\partial \psi^j} \omega = -i\tau_j, \quad \frac{\partial}{\partial \tau_j} \omega = i\psi^j.$$

Therefore

$$\frac{\partial}{\partial \psi^j} \exp \omega = -i\tau_j \exp \omega, \quad \frac{\partial}{\partial \tau_j} \exp \omega = i\psi^j \exp \omega.$$

In analogy to 2) and 3) above we have

Proposition 7.1.1 *For $f \in \wedge V$ we have*

$$F\left(\frac{\partial f}{\partial \psi^j}\right) = i\tau_j F(f) \quad (7.12)$$

and

$$F\left(\psi^j f\right) = -i\frac{\partial F(f)}{\partial \tau_j}. \quad (7.13)$$

Proof. As in the Bosonic case, (7.12) is proved by integration by parts: By linearity, it is enough to check this formula for $f \in \wedge^p V$. We have

$$
\begin{aligned}
F\left(\frac{\partial f}{\partial \psi^j}\right) &= \int \frac{\partial f}{\partial \psi^j} \exp \omega d\psi \\
&= -(-1)^p \int f \frac{\partial \exp \omega}{\partial \psi^j} d\psi \\
&= -(-1)^p \int f(-i\tau_j) \exp \omega d\psi \\
&= i\tau_j \int f \exp \omega d\psi \\
&= i\tau_j F(f).
\end{aligned}
$$

Similarly, (7.13) is verified by differentiating under the integral sign:

$$
\begin{aligned}
-i\frac{\partial F(f)}{\partial \tau_j} &= -i\frac{\partial}{\partial \tau_j} \int f \exp \omega d\psi \\
&= -i \int \frac{\partial}{\partial \tau_j} (f \exp \omega) d\psi \\
&= -i(-1)^p \int f \frac{\partial}{\partial \tau_j} (\exp \omega) d\psi \\
&= -i(-1)^p \int f \cdot i \cdot \psi^j \exp \omega d\psi \\
&= \int \psi^j f \exp \omega d\psi \\
&= F(\psi^j f). \qquad \square
\end{aligned}
$$

Proposition 7.1.2

$$
F^2 = \mathrm{id} \quad \text{if dim } V \text{ is even and} \quad F^2 = i\,\mathrm{id} \quad \text{if dim } V \text{ is odd.} \tag{7.14}
$$

Proof. Let us first verify this formula when applied to the element 1. We have

$$
\begin{aligned}
F1 &= \int \exp(i\tau_k \psi^k) d\psi \\
&= i^d \int \tau_1 \psi^1 \tau_2 \cdots \tau_d \psi^d \\
&= i^d s(d) \tau_1 \cdots \tau_d
\end{aligned}
$$

where $s(d) := (-1)^{\frac{1}{2}d(d-1)}$ is the sign involved in the equation

$$
\tau_1 \psi^1 \cdots \tau_d \psi^d = s(d) \tau_1 \cdots \tau_d \psi^1 \cdots \psi^d.
$$

We have

$$
\begin{aligned}
s(1) &= 1 \\
s(2) &= -1 \\
s(3) &= -1 \\
s(4) &= 1 \\
s(4k+r) &= s(r).
\end{aligned}
$$

So $i^d s(d) = 1$ or i according to whether d is even or odd. Applying the Fourier transform again gives

$$
\begin{aligned}
F(F(1)) &= i^d s(d) F(\tau_1 \cdots \tau_d) \\
&= i^d s(d) \int \tau_1 \cdots \tau_d \exp(-\omega) d\tau \\
&= i^d s(d) \int \tau_1 \cdots \tau_d \\
&= i^d s(d).
\end{aligned}
$$

This proves the formula for $f = 1$. We can now proceed inductively on the degree of f. We have

$$
\begin{aligned}
F^2(\psi^j f) &= F\left(-i \frac{\partial}{\partial \tau_j} F(f)\right) \\
&= \psi^j F^2(f) \\
&= i^d s(d) \psi^j f. \quad \square
\end{aligned}
$$

(We are indebted for this organization of the proof to Matt Leingang.)

Finally, let us compute the Fourier transform of a "Gaussian": Suppose $d = 2m$ and

$$
q = q(\psi^1, \dots, \psi^d) = q_{ij} \psi^i \psi^j, \quad q_{ij} = -q_{ji}, \quad Q = (q_{ij})
$$

as above. We assume that Q is non-degenerate, hence induces an isomorphism of V onto V^*, and therefore a non-degenerate element, call it $q^* \in \wedge^2 V^*$.

Proposition 7.1.3 *The Fourier transform of* $\exp -\frac{q}{2}(\psi^1, \dots, \psi^d)$ *is*

$$
\text{Pfaff } (q) \exp -\frac{q^*}{2} (\tau_1, \dots, \tau_d). \tag{7.15}
$$

Proof. We may assume that we have brought $-\frac{q}{2}$ to the normal form (7.7) an so by iterated integration it suffices to prove (7.15) when $d = 2$ and $-\frac{q}{2} = \lambda \psi^1 \psi^2$. In this case

$$
\exp -\frac{q}{2}(\psi^1, \psi^2) = 1 + \lambda \psi^1 \psi^2
$$

so

$$F\left(\exp -\frac{q}{2}(\psi^1, \psi^2)\right) = \lambda + \tau_1 \tau_2$$

$$= \lambda \left(1 + \frac{1}{\lambda}\tau_1 \tau_2\right)$$

$$= \text{Pfaff}(q)\exp -\frac{q^*}{2}(\tau_1, \tau_2). \quad \square$$

7.2   The Mathai-Quillen Construction

Let V be a d-dimensional vector space over $\mathbf{R}$ equipped with a positive definite inner product and an orientation. Let $\{\psi^1, \ldots \psi^d\}$ be an oriented orthonormal basis of V and $g := o(V)$, the Lie algebra of endomorphisms of V which are skew symmetric with respect to the inner product. We want to consider Fermionic integrals of expressions in $A \otimes \wedge V$, where $A = \Omega_G(V)$. We let $\xi_1, \ldots, \xi_n$ be a basis of g, $n = \frac{1}{2}d(d-1)$. Each $\xi \in g$ is represented on V by a linear transformation M_ξ whose matrix is skew-symmetric in terms of the basis $\psi^1, \ldots, \psi^d$. In other words,

$$\sum_i (M_\xi \psi^i)\psi^i = -\sum_i \psi^i M_\xi \psi^i,$$

since, writing (M_j^i) for the matrix of M_ξ relative to our orthonormal basis, we have

$$\sum_i (M_\xi \psi^i)\psi^i = \sum_{i,j}(M_j^i \psi^j)\psi^i$$

$$= -\sum_{i,j} M_i^j \psi^j \psi^i$$

$$= -\sum_{i,j} \psi^j M_i^j \psi^i$$

$$= -\sum_j \psi^j (M_\xi \psi^j).$$

As usual, we will write M_a for M_{ξ_a} so we write

$$\sum_i (M_a \psi^i)\psi^i = -\sum_i \psi^i M_a \psi^i \quad a = 1, \ldots, n.$$

Let $\{x^1, \ldots, x^n\}$ be the basis of g^* dual to $\{\xi_1, \ldots, \xi_n\}$ and let $\{u_1, \ldots u_d\}$ be the (Bosonic) coordinates on V associated with the basis $\{\psi^1, \ldots, \psi^d\}$ of V. Let $A = \Omega_G(V)$. We will consider Fermionic integration in $A \otimes \wedge V$ followed by the usual (Bosonic) integration over V.

For applications to geometry, we will want to construct an equivariantly closed differential form on V which vanishes rapidly at infinity and whose

(Bosonic) integral over V does not vanish. We will call such a form a **universal Thom form** for reasons which will become apparent in the geometric applications.

Consider the expression

$$\sigma := -\frac{1}{2}\sum_i u_i^2 + i\psi^j du_j - \frac{1}{2}\sum_k \psi^k x^a M_a \psi^k \quad \in S(g^*) \otimes \Omega(V) \otimes \wedge V. \quad (7.16)$$

Up to sign and factors of powers of i and $(2\pi)^{1/2}$, we will find that the Fermionic integral

$$\int \exp \sigma \, d\psi \quad \in \quad S(g^*) \otimes \Omega(V) \quad\quad\quad (7.17)$$

is our desired universal Thom form. We must show that it is d_G-closed and that its (Bosonic) integral over V does not vanish.

We consider the operator

$$d_G = d - x^a \iota_a : \ S(g^*) \otimes \Omega(V) \to S(g^*) \otimes \Omega(V)$$

as acting on $S(g^*) \otimes \Omega(V) \otimes \wedge V$ by letting it act trivially on the last factor, i.e. by acting as $d_G \otimes 1$.

We first compute $d_G\sigma$. Let us consider the two parts of d_G separately. The operator d applies to the u variables in (7.16) yielding

$$d\sigma = -\sum_i u_i du_i.$$

Similarly the operators ι_a only see the second term in (7.16). Since $\iota_a du_k = L_a u_k$ and the ι_a are derivations of odd degree,

$$\begin{aligned}(-x^a \iota_a)\sigma &= ix^a \psi^k L_a u_k \\ &= -ix^a M_a \psi^k u_k,\end{aligned}$$

the last equation stemming from the fact that the u_k are linear coordinates dual to the ψ^i. Since the u_i are even in our superalgebra, we obtain

$$d_G\sigma = -\sum_k u_k \left(du_k + ix^a M_a \psi^k\right).$$

On the other hand

$$\frac{\partial}{\partial \psi^j}\left(-\frac{1}{2}\sum_k \psi^k x^a M_a \psi^k\right) = -x^a M_a \psi^j$$

and

$$\frac{\partial}{\partial \psi^j}\left(i\psi^k du_k\right) = i du_j.$$

Thus

$$\frac{\partial}{\partial\psi^j}\sigma = idu_j - x^a M_a \psi^j$$

so

$$d_G\sigma = \left(\sum_k iu_k \frac{\partial}{\partial\psi^k}\right)\sigma. \tag{7.18}$$

Both sides of (7.18) are derivations (of odd degree) in the algebra $\Omega_G(V)\otimes\wedge V$ applied to the element σ. Since σ is an even element of this commutative superalgebra, any derivation, D (even or odd) satisfies

$$D(\sigma^k) = k\sigma^{k-1}D\sigma$$

and hence

$$D(\exp\sigma) = \exp\sigma D\sigma.$$

So (7.18) implies that

$$d_G(\exp\sigma) = \sum_k iu_k \frac{\partial}{\partial\psi^k}\exp\sigma.$$

Since the derivation d_G on the left of (7.18) does not involve the ψ variables, we can pass it inside the Fermionic integration with respect to ψ to obtain

$$
\begin{aligned}
d_G\int(\exp\sigma)d\psi &= \int d_G(\exp\sigma)d\psi \\
&= i\sum_k u_k \int \frac{\partial}{\partial\psi^k}(\exp\sigma)d\psi \\
&= 0
\end{aligned}
$$

by (7.3). This proves that the form (7.17) is d_G-closed. We now must evaluate its integral over V, and, in particular, show that the value of this integral is not zero.

When we compute this integral, we must extract the coefficient of $du_1\cdots du_d$ in (7.17). We can write (7.17) as

$$e^{-\frac{1}{2}\sum_k u_k^2}\int \exp(i\psi^k du_k)d\psi + \cdots$$

where the remaining terms involve fewer than d factors of the du_k, and can be ignored. The Fermionic integral in this last expression yields

$$i^d e^{-\frac{1}{2}\sum_k u_k^2}(-1)^{\frac{1}{2}d(d+1)}du_1\cdots du_d$$

and so the integral of (7.17) over V is

$$(-i)^d s(d)\int_V e^{-\frac{1}{2}\sum_k u_k^2}du_1\cdots du_d = (-i)^d s(d)(2\pi)^{d/2} \neq 0.$$

If we want the integral to come out to be one we must divide by this non-zero constant. Thus

$$\nu := \frac{i^d}{s(d)(2\pi)^{d/2}} \int \exp \sigma d\psi \tag{7.19}$$

is the universal Thom form as constructed by Mathai and Quillen.

Now suppose that d is even. Let

$$j : \{0\} \to V$$

be the inclusion, and define the normalizing constant γ_d by

$$\gamma_d := (2\pi)^{-\frac{d}{2}}.$$

If we apply j^* to ν, all expressions involving u_i and du_i go to zero, and hence, applying (7.8) to $q = \sum_k \psi^k x^a M_a \psi^k$ we obtain

$$j^*\nu = \gamma_d \mathrm{Pfaff}(x^a M_a). \tag{7.20}$$

In other words, up to the factor γ_d, $j^*\nu$ is the element Pfaff.

The whole discussion above applies to a subgroup K of $O(V)$: If i denotes the inclusion $i : k \to o(V)$ and i^* the dual map from the ring of invariant polynomials on $o(V)$ to $S(k^*)^K$ we obtain

$$j^*\nu = i^*\gamma_d \mathrm{Pfaff}$$

where

$$i^* : S(o(V)^*)^{SO(V)} \to S(k^*)^K.$$

7.3   The Fourier Transform of the Koszul Complex

Let V be a d-dimensional vector space on which G acts in a linear fashion, preserving a positive definite quadratic form

$$u^2 = u_1^2 + \cdots + u_d^2$$

in terms of coordinates associated with an orthonormal basis $\psi^1, \ldots, \psi^d$. Let

$$\Omega_G(V)_e \subset \Omega_G(V)$$

consist of all equivariant differential forms whose coefficients, in terms of the differential forms du_I, are of the form $p(u_1, \ldots, u_d)e^{-u^2/2}$. In other words,

$$\Omega_G(V)_e = (S(g^*) \otimes S(V^*) \otimes \wedge V^*)^G \cdot e^{-u^2/2}$$

where we have identified $\wedge V^*$ with differential forms on V which are linear combinations of the du_I with constant coefficients.

The elements of $\Omega_G(V)_e$ are all integrable and hence we get an integration map

$$\int : H_G(V)_e \to S(g^*)^G.$$

The main goal of this section is to show that this map is a bijection, and, in particular, that

$$H_G^i(V)_e = \begin{cases} 0 & 0 \le i < d \text{ or } i - d \text{ is odd} \\ S^{\frac{i-d}{2}}(g^*)^G & i \ge d \text{ and } i - d \text{ is even} \end{cases} \tag{7.21}$$

Let us first examine this assertion in the non-equivariant case, i.e. where we take $G = \{e\}$ to be trivial and so are considering

$$\Omega(V)_e$$

consisting of all differential forms which are of the form $p^I e^{-u^2/2} du_I$ where the p^I are polynomials on V. We want to think of $\Omega(V)_e$ as a kind of substitute for $\Omega(V)_c$, the space of differential forms of compact support. The analogous theorem for the compactly supported cohomology asserts that $H^i(V)_c$ vanishes for all $i \ne d$, and that $H^d(V)_c = \mathbf{C}$. The standard way of proving this is to identify V with $\mathbf{R}^d$ using the coordinates $(u_1, \ldots, u_d)$ and proving by induction that the fibrations

$$\mathbf{R}^d \to \mathbf{R}^{d-1}, \quad (u_1, \ldots, u_{d-1}, u_d) \mapsto (u_1, \ldots, u_{d-1})$$

induce isomorphisms on cohomology

$$H^k(\mathbf{R}^d)_c \to H^{k-1}(\mathbf{R}^{d-1})_c.$$

See, for example, [BT], page 39. There doesn't appear to be any analogue of this argument in equivariant cohomology which is why we have replaced $\Omega(V)_c$ by $\Omega(V)_e$. With this replacement we can give an alternative argument using the Fourier transform.

So our first order of business will be to prove (7.21) for trivial G. Let $y^1 \ldots, y^d$ be the coordinates in V^* dual to $u_1, \ldots, u_d$ and let $y^2 = (y^1)^2 + \cdots + (y^d)^2$ the dual quadratic form. The ordinary (Bosonic) Fourier transform maps functions of the form

$$p(u)e^{-u^2/2} \in \Omega^0(V)_e$$

into functions of the form

$$\tilde{p}(y)e^{-y^2/2} \in \Omega^0(V^*)_e$$

where $\tilde{p} \in S^*(V)$ is a polynomial in y. We thus get a map

$$F_b : S(V^*) \to S(V), \qquad p \mapsto \tilde{p}.$$

We may also call this the Bosonic Fourier transform by abuse of language. It is the usual Fourier transform where we have suppressed the exponential factor on both sides.

With this suppression of the exponential factor we may identify $\Omega(V)_e$ with $S(V^*) \otimes \wedge(V^*)$ and $\Omega(V^*)_e$ with $S(V) \otimes \wedge V$.

We may now combine the Bosonic and Fermionic Fourier transforms so as to get a super Fourier transform

$$\mathbf{F} := F_b \otimes F_f : \Omega(V)_e \to \Omega(V^*)_e.$$

Under our identifications, the basis elements $\psi^1, \ldots, \psi^d$ of $\wedge^1 V$ are identified with the differential forms $dy^1, \ldots, dy^d$ on V^*. With this in mind, we see that i in (7.9) cancels the $-i$ in (7.13) so that the differential operator

$$d = \frac{\partial}{\partial u_i} \otimes du_i$$

acting on $\Omega(V)_e$ is carried by (conjugation by) $\mathbf{F} = F_b \otimes F_f$ into the operator

$$\delta := y^k \otimes \frac{\partial}{\partial \psi^k}. \tag{7.22}$$

In fact, δ is just the Koszul operator, d_K, of Section 3.1. Let

$$Q := \psi^k \frac{\partial}{\partial y^k}. \tag{7.23}$$

Notice that Q is exactly the operator we introduced in Section 3.1 modulo some changes in notation for the variables. We thus see that

$$[Q, \delta] \tag{7.24}$$

is the derivation given by

$$[Q, \delta] = (k + \ell)\mathrm{id} \quad \text{on} \quad \wedge^k (V) \otimes S^\ell(V)$$

as in Section 3.1. In particular, $(\Omega(V^*)_e, \delta)$ is acyclic, all its cohomology being concentrated in bidegree $(0, 0)$.

Now our super Fourier transform $\mathbf{F}$ carries $S^*(V^*) \otimes \wedge^d(V^*)$ to $S^*(V) \otimes \wedge^0(V)$ sending

$$p\psi^1 \cdots \psi^d \mapsto \tilde{p}$$

and

$$\int p(u)e^{-u^2/2}du_1 \cdots du_d = \tilde{p}(0).$$

In other words, we have shown that (7.21) holds for $G = \{e\}$, the trivial group, and that this isomorphism is realized by integration.

We now show how to modify this argument so as to make it work in the equivariant setting. Let $\xi_1, \ldots, \xi_n$ be a basis of g and $x^1, \ldots, x^n$ the dual basis of $g^* = S^1(g^*)$. Let

$$\xi_a^\sharp = \gamma_{a\ell}^k u_k \frac{\partial}{\partial u_\ell}$$

be the vector field on V corresponding to ξ_a..

The super Fourier transform

$$\mathbf{F} : \Omega(V)_e \to \Omega(V^*)_e$$

extends to a super Fourier transform

$$\mathbf{F} : \Omega_G(V)_e \to \Omega_G(V^*)_e$$

and converts the operator

$$d_G = d - x^a \iota_a$$

into

$$\delta_G = e^{y^2/2} \left(y^a \frac{\partial}{\partial \psi^a} - x^a \gamma_{a\ell}^k \psi^\ell \frac{\partial}{\partial y^k} \right) e^{-y^2/2}. \tag{7.25}$$

To simplify this expression note that

$$\gamma_{a\ell}^k \psi^\ell = -M_a \psi^k.$$

We can write the term inside the parentheses in (7.25) as

$$y^a \frac{\partial}{\partial \psi^a} + x^a M_a \psi^k \frac{\partial}{\partial y^k}$$

and hence the conjugation by $e^{-y^2/2}$ in (7.25) yields

$$\delta_G = y^a \frac{\partial}{\partial \psi^a} - \sum_k x^a M_a \psi^k y^k + x^a M_a \psi^k \frac{\partial}{\partial y^k}. \tag{7.26}$$

Let

$$\beta := \frac{1}{2} \sum_k x^a M_a \psi^k \psi^k \tag{7.27}$$

so that

$$y^a \frac{\partial}{\partial \psi^a} \beta = \sum_k x^a M_a \psi^k y^k.$$

We define

$$\delta_G^\beta := e^{-\beta} \delta_G e^\beta = y^a \frac{\partial}{\partial \psi^a} + \beta^a \frac{\partial}{\partial y^a}. \tag{7.28}$$

where

$$\beta^a := \frac{\partial}{\partial \psi^a} \beta. \tag{7.29}$$

We can regard the derivations Q defined by (7.23) and E defined by (7.24) as being derivations of $\Omega_G(V^*)_e$. Since Q and β^a don't depend on the y^i's, Q supercommutes with $\beta^a \frac{\partial}{\partial y^a}$ and hence

$$[Q, \delta^\beta] = [Q, \delta] = E \tag{7.30}$$

proving the acyclicity of the complex

$$(\Omega_G(V^*)_e, \delta_G^\beta)$$

as a module over $S(g^*)^G$.

This result explains the rather mysterious formula (7.19): The generator of the zero dimensional cohomology group of $(\Omega_G(V^*)_e, \delta_G^\beta)$ is clearly the constant function, 1. Hence the generator of the d-th cohomology group of $(\Omega_G(V)_e, d_G)$ is the inverse super Fourier transform of $\exp(-\frac{y^2}{2} + \beta)$:

$$
\begin{aligned}
\mathbf{F}^{-1}\left(e^{-y^2/2+\beta}\right) &= \mathbf{F}^{-1}\left(e^{-y^2/2}\right)\mathbf{F}_f^{-1}\left(e^\beta\right) \\
&= \frac{e^{-u^2/2}}{i^d s(d)} \int e^\beta e^{idu_k \psi^k}\, d\psi \\
&= \frac{(-1)^d}{i^d s(d)} e^{-u^2/2} \int e^{\beta + i\psi^k du_k}\, d\psi
\end{aligned}
$$

since, by the change of variables $\psi_k \mapsto -\psi_k$ we have

$$idu_k \psi^k \mapsto -idu_k \psi^k = i\psi^k du_k, \quad \beta \mapsto \beta$$

and $(-1)^d$ is the determinant of this change of variables. This is the expression given by (7.19) up to non-zero constants, i.e.

$$\nu := \frac{1}{(2\pi)^{d/2}} \mathbf{F}^{-1}\left(e^{-y^2/2+\beta}\right).$$

7.4   Bibliographical Notes for Chapter 7

1. The role of the Berezin integral within the context of integration on supermanifolds is discussed in Berezin's book [Be]. A nice discussion of its application to physics can be found in the book, "Supermanifolds" by Bryce de Witt [dW].

2. Our treatment of the Fermionic Fourier transform and the super Fourier transform is taken from Kalkman's thesis, [Ka], Section 1.3. The material on the Mathai-Quillen construction of the universal Thom form in

section 7.2 is taken from [MQ] section 6. The alternative construction in Section 7.3, using the super Fourier transform, is closely related to the quantum field theoretic construction of this form by Kalkman in [Ka] section 3.3.

3. Let B be the open unit ball, $\|v\| < 1$, in V and let $\gamma : B \longrightarrow V$ be the map

$$\gamma(v) = \frac{v}{1 - \|v\|^2}.$$

The pull-back by this map of the form (7.19) is an equivariant form on B which vanishes to infinite order at boundary points and hence can be extended to a compactly supported form on all of V by setting it equal to zero on the complement of B. This form is the *compactly supported* version of the universal Mathai-Quillen form (see Section 10.3 for more details).

4. Let M be a compact G-manifold on which G acts freely, and let G act on $M \times V$ by its diagonal action. Let $X = M/G$ and $E = (M \times V)/G$. E is a vector bundle over X with a "typical fiber" V, and X can be embedded in E by identifying it with the zero section. Mathai and Quillen show that one can construct an explicit Thom form representing the cohomology class in $H_c^*(E)$ dual to $[X]$ as follows: Pull the universal Thom form (7.19) back to $M \times V$ by the projection, $M \times V \longrightarrow V$. This gives one an element, ν', of $\Omega_G(M \times V)$. Now apply to ν' the Cartan map

$$\Omega_G(M \times V) \longrightarrow \Omega(E).$$

The image of ν' under this map represents the dual class to $[X]$ in $H_c^*(E)$. (For details see Section 10.4 below.)

5. Using the Mathai-Quillen construction, Atiyah and Jeffrey show that Witten's formula for the Donaldson invariants of a four-manifold has a beautiful interpretation in terms of Euler numbers of (infinite dimensional) vector bundles. (See [AJ].) Their basic observation, is that the construction described in Section 4, when appropriately interpreted, even makes sense when V is infinite-dimensional!

Chapter 8

Characteristic Classes

Recall from section 4.5 that if A is a G^* module, then we have a characteristic homomorphism

$$\kappa_* : S(g^*)^G \to H_G(A),$$

and that the elements of the image of κ_* are known as characteristic classes. But we have not really written down what the ring $S(g^*)^G$ is for any group G. The main function of this chapter is to remedy this by summarizing standard computations of $S(g^*)^G$ for various important groups. Suppose that $\phi : K \to G$ is a Lie group homomorphism, and let k denote the Lie algebra of K. The induced Lie algebra map $k \to g$ dualizes to a map $g^* \to k^*$ which extends to an algebra homomorphism $S(g^*)^G \to S(k^*)^K$. We will examine this homomorphism for various examples of inclusions of classical groups.

In important geometric applications, we want to apply the notion of characteristic classes to the the study of vector bundles. So we begin this chapter with a review of standard geometrical constructions which motivate the choice of groups and inclusions we study.

8.1 Vector Bundles

Let $E \to X$ be a complex vector bundle. Choose a Hermitian structure on E and let $M = \mathcal{F}(E)$ denote its bundle of unitary frames. So a point of M consists of a pair $(x, \mathbf{e})$ where $x \in X$ and $\mathbf{e} = (e_1, \dots, e_n)$ is an orthonormal basis of E_x. The group $G = U(n)$ acts on the right, where

$$A \in G : \quad (x, \mathbf{e}) \mapsto (x, \mathbf{e}A).$$

This makes M into a principal G-bundle over X and hence we get a map

$$\kappa : S(g^*)^G \to H^*(X) = H_G^*(M).$$

The elements of the image of this map gives a subring of the cohomology ring of X called the ring of characteristic classes of the vector bundle E. Its definition depended on the choice of a Hermitian metric. Let us show that the functorial properties of equivariant cohomology imply that κ does not depend on this choice so the terminology "characteristic" is justified.

So let h_0 and h_1 be two choices of Hermitian structures. Then for every $t \in [0, 1] = I$

$$h_t := (1 - t)h_0 + th_1$$

is a Hermitian structure. Let $\tilde{M} = \tilde{\mathcal{F}}(E)$ denote the bundle over $X \times I$ whose fiber over (x, t) consists of all frames, e of E_x which are orthonormal with respect to h_t. It is a principal G bundle, and we have the injections of M_0 and M_1 into $\tilde{M}$ setting $t = 0$ or $t = 1$ where M_0 is the frame bundle associated to h_0 and M_1 the frame bundle associated to h_1. Let j_0 and j_1 denote the corresponding injections of $X \to X \times I$, so $j_0(x) = (x, 0)$, $j_1(x) = (x, 1)$. Then functoriality implies that

$$\kappa_0 = j_0^* \tilde{\kappa}, \qquad \kappa_1 = j_1^* \tilde{\kappa}$$

in the obvious notation. But since j_0 and j_1 are homotopic they induce the same homomorphism on cohomology.

In the above discussion we could have taken E to be a real vector bundle, h a real scalar product, and M to be the bundle of orthonormal frames. The group G is then the orthogonal group $O(n)$. Similarly, we could have taken E to be an oriented real vector bundle, h a real scalar product, M to be the bundle of oriented orthonormal frames and $G = SO(n)$ the special orthogonal group. In all three cases, $U(n)$, $O(n)$, $SO(n)$ we must examine the ring $S(g^*)^G$ whose structure we will recall below. We state these results and refer to standard references such as [Chev] for the proofs. In each case there are standard generators, whose images under κ are called Chern classes for the case of $U(n)$, Pontryagin classes for the case of $O(n)$ and one class, called the Pfaffian, in addition to the Pontryagin classes in the case of $SO(2n)$.

We will also examine the case of a symplectic vector bundle. We will find there are standard ways of putting a complex structure on such a vector bundle, and that two such ways differ by a homotopy. Therefore the characteristic classes associated to these choices of complex structures agree.

8.2   The Invariants

8.2.1 $G = U(n)$

We may identify the Lie algebra of $U(n)$ with space of all matrices of the form iA where A is self adjoint, and hence (using the trace pairing and forgetting about the inessential i) may identify g^* with the space of self adjoint matrices with the coadjoint action being conjugation:

$$U : A \mapsto UAU^{-1}, \quad U \in U(n).$$

Define the polynomial c_i, of degree i in A to be the coefficient of $(-1)^i \lambda^{n-i}$ in the characteristic polynomial of A:

$$\det(\lambda - A) = \lambda^n - c_1(A)\lambda^{n-1} + \cdots + (-1)^n c_n(A).$$

For instance, $c_1(A) = \operatorname{tr} A$ and $c_n(A) = \det A$. The polynomials c_i are clearly invariant under the adjoint representation, as the characteristic polynomial is. It is a theorem that they generate the ring of invariants.

The characteristic classes corresponding to the c_i for a complex vector bundle are called its **Chern classes**.

8.2.2 $G = O(n)$

We may identify g and hence g^* with the space of skew adjoint matrices. For such a matrix we have

$$\det(\lambda - A) = \det(\lambda - A^t) = \det(\lambda + A)$$

so all the coefficients of λ^{n-i} in the characteristic polynomial of A vanish when i is odd. We may write

$$\det(\lambda - A) = \lambda^n + p_1(A)\lambda^{n-2} + p_2(A)\lambda^{n-4} + \cdots.$$

The polynomials p_i of degree $2i$ generate the ring $S(g^*)^G$. The corresponding characteristic classes for a real vector bundle are called its **Pontryagin classes**.

8.2.3 $G = SO(2n)$

The Lie algebra g is the same as for $O(2n)$ so all the p_i are invariant polynomials. There is one additional invariant which is not a polynomial in the p_i called the **Pfaffian**. It is defined as follows: To each $A \in g$ and $v, w \in V = \mathbf{R}^{2n}$ set

$$\omega_A(v, w) := (Av, w)$$

where $(\, , \,)$ denotes the scalar product. We have

$$\omega_A(w, v) = (v, Aw) = -(Av, w) = -\omega_A(v, w)$$

so $\omega_A \in \wedge^2(V^*)$ is an alternating bilinear form and the map $A \mapsto \omega_A$ is a linear isomorphism. The element

$$\frac{1}{n!}\omega_A^n$$

is an element of $\wedge^{2n}(V^*)$ and the map $A \mapsto \frac{1}{n!}\omega_A^n$ depends only on the scalar product and so is $O(2n)$ invariant. However the group $SO(2n)$ preserves a basis element, vol, of $\wedge^{2n}(V^*)$ where

$$\mathrm{vol} := e_1^* \wedge e_2^* \wedge \cdots \wedge e_{2n}^*.$$

Here $e_1, e_2, \ldots, e_{2n}$ is any oriented orthonormal basis and $e_1^*, \ldots, e_{2n}^*$ the dual basis. We may then define Pfaff(A) by

$$\frac{1}{n!}\omega_A^n = \text{Pfaff } (A)\text{vol}. \tag{8.1}$$

It is a polynomial function of A of degree n which is $SO(2n)$ invariant. For any A we can find an oriented orthonormal basis relative to which the matrix A takes the form

$$\begin{pmatrix} \begin{pmatrix} 0 & \lambda_1 \\ -\lambda_1 & 0 \end{pmatrix} & 0 & \cdots & 0 \\ 0 & \begin{pmatrix} 0 & \lambda_2 \\ -\lambda_2 & 0 \end{pmatrix} & \cdots & 0 \\ \vdots & \vdots & \ddots & \vdots \\ 0 & 0 & \cdots & \begin{pmatrix} 0 & \lambda_n \\ -\lambda_n & 0 \end{pmatrix} \end{pmatrix}.$$

Relative to this orthonormal basis we have

$$\omega_A = \lambda_1 e_1^* \wedge e_2^* + \cdots + \lambda_n e_{2n-1}^* \wedge e_{2n}^*$$

so

$$\frac{1}{n!}\omega_A^n = \lambda_1 \cdots \lambda_n \text{vol}$$

and hence

$$\text{Pfaff } (A) = \lambda_1 \cdots \lambda_n.$$

On the other hand

$$\det(A) = \lambda_1^2 \cdots \lambda_n^2.$$

So we see that we have the general formula

$$\text{Pfaff }^2 = \det. \tag{8.2}$$

Thus the square of $(2\pi)^{-n}$ Pfaff is equal to p_n. Notice that for odd dimensions the determinant of any antisymmetric matrix vanishes as we have seen above, so this phenomenon does not occur.

The characteristic class corresponding to the Pfaffian for a real, 2n dimensional, oriented vector bundle E is called the **Euler class** of the vector bundle and denoted by

$$e(E) \in H^{2n}(X).$$

8.3   Relations Between the Invariants

We can regard $O(n)$ as the subgroup of $U(n)$ consisting of real matrices: $U(n)$ consists of all complex matrices satisfying

$$MM^* = I \quad \text{where} \quad M^* = \bar{M}^t.$$

If $M = \bar{M}$ is a real matrix then this becomes

$$MM^t = I$$

which is the condition for a real matrix to be orthogonal. The inclusion $o(n) \to u(n)$ induces a homomorphism $S(u(n)^*)^{U(n)} \to S(o(n)^*)^{O(n)}$. We will check below what the images of the c_i are under this restriction map.

Any complex vector space of dimension n can be regarded as a real vector space dimension $2n$. The complex structure induces an orientation on this real vector space. If the complex vector space has a Hermitian structure, the real part of this structure gives a real scalar product. These two facts combine to give an embedding $U(n) \to SO(2n)$. We will also examine the restriction map $S(o(2n)^*)^{SO(2n)} \to S(u(n)^*)^{U(n)}$ corresponding to this embedding.

8.3.1 Restriction from $U(n)$ to $O(n)$

We have identified the Lie algebra $u(n)$ with the self-adjoint matrices, where A generates the one parameter group $\exp itA$. For these group elements to be real matrices, we must have

$$\overline{\exp itA} = \exp(-it\bar{A}) = \exp itA$$

so

$$\bar{A} = -A$$

or

$$A = iB$$

where B is real and

$$B^t = -B.$$

We get

$$
\begin{aligned}
\det(\lambda - iB) &= \lambda^n - \lambda^{n-1}c_1(iB) + \lambda^{n-2}c_2(iB) - \cdots \\
&= \lambda^n - i\lambda^{n-1}c_1(B) + i^2\lambda^{n-2}c_2(B) - \cdots \\
&= \lambda^n + \lambda^{n-2}p_1(B) + \cdots
\end{aligned}
$$

so under the restriction map we get

$$c_{2i+1} \mapsto 0, \qquad c_{2i} \mapsto (-1)^i p_i. \tag{8.3}$$

This has the following consequence for characteristic classes: Let $E \to X$ be a real vector bundle and $E \otimes \mathbf{C}$ its complexification. The Chern classes of this complexified bundle are related to the Pontryagin classes of the original bundle by

$$c_{2i+1}(E \otimes \mathbf{C}) = 0, \quad c_{2i}(E \otimes \mathbf{C}) = (-1)^i p_i(E). \tag{8.4}$$

8.3.2 Restriction from $SO(2n)$ to $U(n)$

We recall two facts:

a) Let V be a complex vector space and $C : V \to V$ a complex linear transformation. We can regard V as a real vector space (of twice the dimension) and C as a real linear transformation. The relation between the determinant of C regarded as a real linear transformation and its determinant regarded as a complex linear transformation is given by

$$\det{}_{\mathbf{R}}(C) = |\det{}_{\mathbf{C}}(C)|^2 = \det{}_{\mathbf{C}} C \cdot \overline{\det{}_{\mathbf{C}} C}.$$

b) If A is a self-adjoint transformation (relative to some Hermitian form) on a complex vector space then its determinant (over $\mathbf{C}$) is real. Indeed, we may diagonalize any such operator and its eigenvalues are real. By the same argument, the coefficients of the characteristic polynomial

$$\det(\lambda - A) = \lambda^n - c_1(A)\lambda^{n-1} + c_2(A)\lambda^{n-2} - \cdots$$

are all real.

The subalgebra $u(n) \subset o(2n)$ consists of complex linear transformations B where $A = iB$ is self-adjoint. Hence $\det_{\mathbf{R}}(\lambda - B) =$

$$\left[\lambda^n - (-i)c_1(A)\lambda^{n-1} + (-i)^2\lambda^{n-2}c_2(A) - \cdots\right] \cdot \left[\lambda^n - ic_1(A)\lambda^{n-1} + \cdots\right]$$

so the restriction map in question is given by

$$p_k \mapsto \sum_{j=0}^{2k}(-1)^{k-j}c_{2k-j}c_j \tag{8.5}$$

where $c_0 = 1$ and, of course, $c_m = 0$, $m > n$. In particular

$$p_n \mapsto c_n^2,$$

and hence (we may choose the orientation so that)

$$\text{Pfaff} \mapsto c_n. \tag{8.6}$$

Applied to a complex vector bundle regarded as an oriented real vector bundle we get the the corresponding equations relating the Pontryagin classes to the Chern classes. In particular, the Euler class is given by

$$e(E) = (2\pi)^{-n}c_n(E). \tag{8.7}$$

8.3.3 Restriction from $U(n)$ to $U(k) \times U(\ell)$

Consider a self-adjoint operator of the form $A = A_1 \oplus A_2$ relative to the direct sum decomposition $\mathbf{C}^n = \mathbf{C}^k \oplus \mathbf{C}^\ell$ where $k + \ell = n$. Then

$$\det(\lambda - A) = \det(\lambda - A_1) \cdot \det(\lambda - A_2).$$

So if we use the notation $c_{j,p}$ to denote the Chern polynomials associated to $U(p)$, the restriction map from $S(u(n)^*)^{U(n)} \to S(u(k)^*)^{U(k)} \otimes S(u(\ell)^*)^{U(\ell)}$ is given by

$$c_{j,n} \mapsto \sum_{r+s=j} c_{r,k} c_{s,\ell}. \tag{8.8}$$

Applied to complex vector bundles we get: Let $E = E_1 \oplus E_2$ be a decomposition of a vector bundle of rank n into a direct sum of vector bundles of rank k and ℓ. Then

$$c_j(E) = \sum_{r+s=j} c_r(E_1) c_s(E_2).$$

We can write this more succinctly as follows. For any complex vector bundle, E, define its **total Chern class** as

$$c(E) = a + c_1(E) + \cdots + c_n(e).$$

Then

$$c(E) = c(E_1) c(E_2) \tag{8.9}$$

when

$$E = E_1 \oplus E_2.$$

Using the relation between Chern classes of the complexification of a real vector bundle and the Pontryagin classes given above, (8.4), we get an analogous formula for the Pontryagin classes.

Equation (8.9) generalizes in the obvious way when E is decomposed into a direct sum of several vector bundles. At the extreme, suppose that E splits as a direct sum of line bundles,

$$E = L_1 \oplus L_2 \oplus \cdots \oplus L_n.$$

Then

$$c(E) = (1 + c_1(L_1))(1 + c_1(L_2)) \cdots (1 + c_1(L_n))$$

so

$$c_k(E) = \sigma_k(c_1(L_1), \ldots, c_1(L_n)) \tag{8.10}$$

where σ_k denotes the k-th elementary symmetric function.

8.4   Symplectic Vector Bundles

8.4.1   Consistent Complex Structures

Let V be a real finite dimensional vector space and let $(\ ,\ )_0$ be a positive definite scalar product on V. We let $\mathcal{A}$ denote the space of linear transformations which are self-adjoint relative to this scalar product, so satisfy

$$(Au, v)_0 = (u, Av)_0 \quad \forall\ u, v \in V.$$

We let $\mathcal{P}$ denote the open subset of $\mathcal{A}$ consisting of all positive definite self-adjoint linear transformations, so $A \in \mathcal{P}$ if and only if

$$(Au, u)_0 > 0 \ \forall \, u \neq 0.$$

Lemma 8.4.1 *The map* $Sq : A \mapsto A^2$ *is a diffeomorphism of* $\mathcal{P}$ *onto itself.*

Proof. Let $e_1, \ldots, e_n$ be an orthonormal basis of eigenvalues of A with eigenvalues $\lambda_1, \ldots, \lambda_n$ so all the $\lambda_i > 0$. Then any $A \in \mathcal{P}$ has a unique positive definite square root, namely the operator with the same eigenvectors and with eigenvalues $\sqrt{\lambda_1}, \ldots, \sqrt{\lambda_n}$ where we take the positive square roots. This shows that the map Sq is bijective. We must show that it is a diffeomorphism. The tangent space to $\mathcal{P}$ is $\mathcal{A}$ and we must show that for any $A \in \mathcal{P}$ the map

$$d(Sq)_A : \mathcal{A} \to \mathcal{A}$$

is injective. We have

$$d(Sq)_A(W) = AW + WA, \quad W \in \mathcal{A}.$$

Suppose the right hand side of this equation were 0. We would then get

$$AWe_i = -\lambda_i We_i \ \forall \, i.$$

If $We_i \neq 0$ it would be an eigenvector of A with a negative eigenvalue which is impossible. Hence $W = 0$. $\square$

As a corollary we conclude that the inverse map

$$A \mapsto \sqrt{A}$$

is smooth.

Suppose that ω is a symplectic form on V (so V is even dimensional and ω is a non-degenerate antisymmetric bilinear form). Then there is a unique anti-symmetric linear operator, $B : V \to V$ such that

$$\omega(u, v) = (u, Bv)_0 \ \forall \, u, v \in V.$$

Notice for future use that B depends linearly on ω and smoothly on the choice of scalar product, $(\ , \ )_0$.

The operator $B^t B = -B^2$ is positive definite and and has a unique square root, C which depends smoothly on B and hence on ω. So

$$C := \sqrt{-B^2}.$$

Let

$$J := BC^{-1}. \tag{8.11}$$

Since B and C commute we have

$$J^2 = -I \tag{8.12}$$
$$CJ = JC \tag{8.13}$$
$$B = JC. \tag{8.14}$$

We have

$$\omega(u,v) = (u, Bv)_0 = (u, CJv)_0$$

so if we define a new scalar product $(\ ,\ )$ by

$$(u,v) := (u, Cv)_0$$

we have

$$\omega(u,v) = (u, Jv) \tag{8.15}$$

so

$$(Ju, v) = -(u, Jv) \tag{8.16}$$

and

$$(u,v) = \omega(Ju, v). \tag{8.17}$$

We may think of J as defining a complex structure on V and then may define

$$h(u,v) = (u,v) + i\omega(u,v). \tag{8.18}$$

It is easy to check that h is a Hermitian form relative to J, that is that

$$h(v,u) = \overline{h(u,v)}$$
$$h(Ju, v) = ih(u,v)$$
$$h(u,u) > 0 \text{ if } u \neq 0.$$

Notice that J and h depend smoothly on $(\ ,\ )_0$ and ω and if ω and $(\ ,\ )_0$ are invariant under the action of some group G then so are J and h.

8.4.2 Characteristic Classes of Symplectic Vector Bundles

Let $E \to X$ be a symplectic vector bundle. This means that each fiber E_x has a symplectic form which varies smoothly in the usual sense. We may put a scalar product on this vector bundle which then determines a complex structure and an Hermitian structure (depending on our choice of scalar product). A homotopy between two different choices of scalar product induces a homotopy between the corresponding complex and Hermitian structures. So the characteristic classes associated to the corresponding unitary frame bundles are the same. In this way the Chern classes (for any choice of real scalar product) are invariants of the symplectic structure.

8.5   Equivariant Characteristic Classes

Let K and G be compact Lie groups and set $H := K \times G$. If M is an H-manifold, we can regard M as a K-manifold on which G acts, the G action commuting with the K action. If the K action is free, one gets an induced action of G on the quotient manifold

$$X := M/K,$$

and, by (4.28), a Chern-Weil map

$$\kappa_K : S(k^*)^K \to H_G(X). \tag{8.19}$$

The elements of the image of this map are called the *equivariant* characteristic classes. Here are some important examples:

8.5.1   Equivariant Chern classes

Let X be a G-manifold and $E \to X$ a complex rank n vector bundle on which G acts as vector bundle automorphisms. Thus, if $x \in X$ and $a \in G$, the action of a on E maps the fiber of E over x linearly onto the fiber over ax. If we equip E with a G-invariant Hermitian inner product, we get a G-action on the associated unitary frame bundle $\mathcal{F}(E)$ which commutes with the action of $U(n)$ as described in Section 8.1. Thus, if we take $K = U(n)$ we get from (8.19) a map from $S(k^*)^K$ to $H_G(X)$. The images under this map of the elements c_i described in subsection 8.2.1 will be called the **equivariant Chern classes**. Just as in Section 8.1 one can prove that they are independent of the choice of Hermitian inner product.

If the vector bundle $E \to X$ is real, we can define the equivariant versions of the Pontryagin classes by the same method, and if E is an oriented real vector bundle of even rank we get an equivariant Euler class — either by mimicking the construction above with $K = SO(n)$ or, if E has an underlying complex structure, by defining the equivariant Euler class in terms of the equivariant Chern classes as in 8.7. These classes satisfy the same identities as those described in Section 8.3, e.g. the identities (8.4), (8.7), and (8.9).

8.5.2   Equivariant Characteristic Classes of a Vector Bundle Over a Point

If the vector bundle $E \to X$ is topologically trivial, its characteristic classes vanish. But this need not be true of its equivariant characteristic classes. For example, consider a vector bundle over a point $E \to$ pt. This is just an ordinary n-dimensional vector space on which G acts as linear automorphisms. Equipping E with a G-invariant metric, we can regard the representation of G on E as a homomorphism

$$G \to K \sim U(n).$$

This gives a homomorphism of the rings of invariants:

$$S(k^*)^K \to S(g^*)^G = H_G(\text{pt.}),$$

and the equivariant Chern classes are just the images of the $c_i \in S(k^*)^K$.

8.5.3 Equivariant Characteristic Classes as Fixed Point Data

Suppose that G has a positive dimensional center. In other words, suppose that the circle group S^1 sits inside G as a central subgroup. Let X be a G-manifold, Y a connected component of X^{S^1}, and $E \to Y$ the normal bundle of Y in X. Since the action of S^1 commutes with the action of G, we get an action of G on E as vector bundle automorphisms and an action of S^1 which commutes with this G-action. We claim that this S^1-action endows E with a complex structure which is preserved by G. For this we'll need the following lemma:

Lemma 8.5.1 *Let V be a vector space over $\mathbf{R}$ and $\rho : S^1 \to GL(V)$ a representation of S^1 on V which leaves no vector fixed except 0. Let*

$$A = \frac{d}{dt}\rho(e^{it})|_{t=0}.$$

Then there exists a unique decomposition:

$$V = V_1 \oplus \cdots \oplus V_k \qquad (8.20)$$

and positive integers, $0 < m_1 < \cdots < m_k$ such that $A = m_i J_i$ on V_i with $J_i^2 = -I$.

Proof. Equip V with an S^1-invariant inner product, i.e., an inner product satisfying:

$$(\rho(e^{it})v, \rho(e^{it})w) = (v, w).$$

Differentiating and setting $t = 0$,

$$(Av, w) + (v, Aw) = 0$$

i.e., $A^t = -A$. Since $V^{S^1} = \{0\}$, A is invertible and hence $A^t A$ is positive definite. Let $0 < \lambda_1 < \cdots < \lambda_k$ be the distinct eigenvalues of $A^t A$ and

$$V = V_1 \oplus \cdots \oplus V_k$$

the decomposition of V into the eigenspaces corresponding to these eigenvalues. On V_i, $A^2 = -\lambda_i I$, so $A_i = m_i J_i$, with $m_i = \sqrt{\lambda_i}$ and $J_i^2 = -I$. The eigenvalues of J_i are $\pm\sqrt{-1}$ so the eigenvalues of $\exp(2\pi A)$ on V_i are $\exp(\pm 2\pi m_i \sqrt{-1})$. However, $\exp 2\pi A = I$; so the m_i's are integers. The uniqueness of the decomposition (8.20) follows from the fact that V_i is the kernel of $A^2 + m_i^2 I$. $\square$

Corollary 8.5.1 V *admits a canonical complex structure. In particular* V *is even dimensional and has a canonical orientation.*

Proof.   A canonical complex structure on V is defined by $J = J_1 \oplus \cdots \oplus J_k$. $\square$

Now, since every $y \in Y$ is fixed by S^1, we get a representation of S^1 on E_y with no trivial component. So we can decompose E_y into S^1-invariant subspaces

$$E_y = \bigoplus E_{y,k}$$

where the action of S^1 on $E_{y,k}$ is given by

$$t \mapsto \exp(a_k t J_{y,k}), \quad t \in \mathbf{R}/2\pi\mathbf{Z}$$

where the a_k are positive integers, $a_j \neq a_k$ for $j \neq k$ and

$$J_{y,k}^2 = -I.$$

The $E_{y,k}$ and $J_{y,k}$ depend smoothly on y and so define a canonical decomposition

$$E = \bigoplus E_k$$

of E into complex vector bundles.

The equivariant Chern classes of these vector bundles are important topological invariants of X with its G-action. In particular, the equivariant Euler class defined by (8.7) will play a fundamental role in the "localization theorem" which we will discuss in Chapter 10.

8.6   The Splitting Principle in Topology

Let X be a manifold and let $E \to X$ a complex rank n vector bundle over X. Given a manifold Y and a smooth map $\gamma : Y \to X$, the **pull-back** $\gamma^* E$ of E to Y is defined to be the set

$$\gamma^* E := \{(y, e), \ y \in Y, e \in E_{\gamma(y)}\}$$

with the obvious projection (onto the first factor). The splitting principle (cf.[BT] Section 21) asserts:

Theorem 8.6.1 *For every vector bundle* $E \to X$ *there exists a manifold* Y *and a fibration* $\pi : Y \to X$ *such that*

1. $\pi^* : H^*(X) \to H^*(Y)$ *is injective, and*

2. $\pi^* E$ *splits into a direct sum of line bundles.*

A manifold Y with these properties is called a **splitting manifold** for $E \rightarrow X$. We will deduce the existence of such a manifold from the abstract splitting principle which we established in section 6.8. As in section 8.1, let $\mathcal{F}(E)$ be the unitary frame bundle of E relative to a choice of Hermitian metric. Let

$$
\begin{aligned}
M &:= \mathcal{F}(E), \\
T &:= \text{a Cartan subgroup of } U(n), \\
Y &:= M/T.
\end{aligned}
$$

Since we may identify X with $M/U(n)$, we get a fibration

$$\pi : Y \rightarrow X$$

and Theorem 6.8.3 establishes property 1) in our theorem. Furthermore, not only is $\pi^* : H^*(X) \rightarrow H^*(Y)$ injective, we have the identification

$$\pi^* H^*(X) \cong H^*(Y)^{\mathbf{W}} \tag{8.21}$$

where $\mathbf{W}$ is the Weyl group. We now show that Y is a splitting manifold. For $U(n)$, We may choose T to be the group of diagonal unitary matrices, so two orthonormal frames $\mathbf{e}$, $\mathbf{e}' \in \mathcal{F}(E)$ over $p \in X$

$$\mathbf{e} = (e_1, \ldots, e_n), \qquad \mathbf{e}' = (e_1', \ldots, e_n')$$

lie in the same T-orbit if and only if there exist $\{\theta_k\}$ such that

$$e_k' = e^{i\theta_k} e_k, \quad 0 \le \theta_k < 2\pi, \tag{8.22}$$

i.e. if and only if e_k and e_k' span the same one dimensional subspace $L_p(k) \subset E_p$. Thus a T-orbit in $\mathcal{F}(E)_p$ defines a decomposition,

$$E_p = L_1(p) \oplus \cdots \oplus L_n(p) \tag{8.23}$$

into mutually orthogonal one-dimensional subspaces. Conversely, given such a decomposition, choosing a unit vector in each summand gives an orthonormal frame, $\mathbf{e}$, and two different choices differ by a transformation of the form (8.22). So the decomposition (8.23) defines a T-orbit in $\mathcal{F}(E)$. We have thus proved

Proposition 8.6.1 *There is a one-to-one correspondence between orthogonal decompositions (8.23) and points on the fiber of Y over p.*

From this description of Y we see that

$$\pi^* E = L_1 \oplus \cdots \oplus L_n \tag{8.24}$$

where the L_l are the "tautological" line bundles associated with the splitting (8.23). This establishes property 2) and proves the theorem. □

The Weyl group $\mathbf{W}$ is just the group of permutations of $\{1, \ldots, n\}$; it acts on Y by permuting the summands in (8.23). In other words, $\tau \in \mathbf{W} = S_n$ sends the point of Y represented by the decomposition (8.23) into the point represented by the decomposition

$$E_p = L_{\tau^{-1}(1)} \oplus \cdots \cdots L_{\tau^{-1}(n)}.$$

In particular.

$$\tau^* L_k = L_{\tau(k)}. \tag{8.25}$$

Let $c(L_k)$ denote the Chern class of L_k. It follows that

$$\tau^* c(L_k) = c(L_{\tau(k)}). \tag{8.26}$$

By Theorem 6.8.4, the $c(L_k)$ generate $H^*(Y)$ as a ring over $H^*(X)$ and so (8.26) specifies the action of $\mathbf{W}$ on $H^*(Y)$. As an independent confirmation of (8.21), we note that by (8.10),

$$\pi^* c_k(E) = \sigma_k(c_1(L_1), \ldots, c_1(L_n)) \tag{8.27}$$

where σ_k is the k-th symmetric function. By the basic theorem of symmetric functions, ([VdW] page 78), *every* symmetric polynomial is a polynomial function of $\sigma_1, \ldots, \sigma_n$. Hence, by (8.26) and (8.27) every $\mathbf{W}$-invariant of $H^*(Y)$ is in $\pi^* H^*(X)$.

If X is a G-manifold and $E \to X$ is a vector bundle on which G acts by vector bundle automorphisms, there is an equivariant version of the splitting principle: As we pointed out in Section 8.5.1, the action of G on E lifts to an action of G on $\mathcal{F}(E)$ which commutes with the action of $U(n)$. Hence one gets an action of G on the quotient $Y = \mathcal{F}(E)/T$. The fibration $\pi : Y \to X$ becomes an equivariant fibration and hence induces a map

$$\pi^* : H_G(X) \to H_G(Y).$$

By Theorem 6.8.2 this map is injective . It is also clear that splitting (8.24) is an equivariant splitting. So we have proved

Theorem 8.6.2 *There exists an equivariant splitting manifold of E. That is, there exists a G-manifold Y and a G-fibration $\pi : Y \to X$ such that*

1. $\pi^* : H_G(X) \to H_G(Y)$ *is injective, and*

2. $\pi^* E$ *splits equivariantly into a direct sum of G-line bundles.*

8.7   Bibliographical Notes for Chapter 8

1. Most of the material in this section is fairly standard. For a more detailed treatment see [MS] sections 14-15, or [BT] chapter IV.

2. The characteristic classes of a symplectic vector bundle have generated a lot of interest lately because of their role in the proof of the "quantization commutes with reduction" theorem and its many variants. See Meinrenken, [Me1] and [Me2], Vergne [Ve] and Duistermaat-Guillemin-Meinrenken-Wu [DGMW].

3. To prove that for $G = U(n)$ the c_i's generate $S(g^*)^G$, we note that by Chevalley's theorem (see section 6.8) the map

$$S(g^*)^G \to S(t^*)^{\mathbf{W}}$$

is bijective (t being the Lie algebra of the Cartan subgroup, T, of G and W being the Weyl group). For $G = U(n)$, T is the group of diagonal unitary matrices, i.e.,

$$T = S^1 \times \cdots \times S^1 \qquad (n \text{ copies})$$

and $t = \sqrt{-1}\mathbf{R}^n$. Moreover, W is the group, Σ_n, of permutations of the set $\{1, 2, \cdots, n\}$ and acts on t by permuting the coordinates, $(x_1, \cdots, x_n)$ of $x \in \sqrt{-1}\mathbf{R}^n$. Thus

$$S(g^*)^G \cong \mathbf{C}[x_1, \cdots, x_n]^{\Sigma_n}$$

and under this identification the c_i's go into the elementary symmetric polynomials in $x_1, \cdots, x_n$. Hence to prove that the c_i generate $S(g^*)^G$ it suffices to show that the elementary symmetric polynomials generate the ring $\mathbf{C}[x_1, \cdots, x_n]^{\Sigma_n}$. For a proof of this see, for instance, [VdW] section 26.

4. Our assertion that the p_i's generate $S(g^*)^G$ when $G = O(n)$ can be proved by a similar argument. Let $\mathbf{W}$ be the semi-direct product

$$\mathbf{W} = \Sigma_n \triangleright \mathbf{Z}_2^n$$

and let $\mathbf{W}$ act on $\mathbf{C}[x_i, \cdots, x_n]$ by letting

$$(\sigma, \epsilon)(x_1, \cdots, x_n) = (\epsilon_1 x_{\sigma(1)}, \cdots, \epsilon_n x_{\sigma(n)}),$$

σ being a permutation of $\{1, 2, \cdots, n\}$ and $\epsilon = (\pm 1, \cdots, \pm 1)$ an element of $\mathbf{Z}_2^n$. To prove that the p_i's generate $S(g^*)^G$ it suffices to show that $\mathbf{C}[x_1, \cdots, x_n]^{\mathbf{W}}$ is generated by the elementary symmetric functions in $x_1^2, \cdots, x_n^2$ and this follows easily from the results in [VdW], Section 26, that we cited above.

5. An alternative description of the Pfaffian is the description in section 7.1.3 as the Gaussian integral (7.6).

6. For a different approach to the theory of equivariant characteristic classes see [BGV] Section 7.1. (However, their approach also involves "super" ideas: in particular, superconnections on vector bundles).

7. Bott and Tu [BT] give the following pragmatic formulation of the splitting principle:

> "To prove a polynomial identity in the Chern classes of complex vector bundles it suffices to prove it under the assumption that the vector bundles are sums of line bundles."

In addition, if the vector bundles are G-vector bundles the same is true of the *equivariant* Chern classes.

Chapter 9

Equivariant Symplectic Forms

9.1 Equivariantly Closed Two-Forms

Suppose we are given an action $G \times M \to M$. In the Cartan model, an element

$$\tilde{\omega} \in \Omega^2_G(M) = \left(\Omega^2(M)^G \otimes S^0(g^*) \right) \oplus \left(\Omega^0(M) \otimes S^1(g^*) \right)^G$$

can be written as

$$\tilde{\omega} = \omega - \phi$$

where $\omega \in \Omega^2(M)$ is a two-form invariant under G and $\phi \in \left(\Omega^0(M) \otimes g^* \right)^G$ can be considered as a G equivariant map,

$$\phi : g \to \Omega^0(M) = \mathcal{F}(M)$$

from the Lie algebra, g to the space of smooth functions on M. For each $\xi \in g$, $\phi(\xi)$ is a smooth function on M, and this function depends linearly on ξ. Therefore, for each $m \in M$, the value $\phi(\xi)(m)$ depends linearly on ξ, so we can think of ϕ as defining a map from M to the dual space g^* of the Lie algebra of g:

$$\phi : M \to g^*, \quad \langle \phi(m), \xi \rangle := \phi(\xi)(m).$$

We will also use the notation

$$\phi^\xi \quad \text{for} \quad \phi(\xi).$$

The condition that $\tilde{\omega}$ be equivariantly closed now translates into two conditions, $d\omega = 0$ and $-\iota_\xi \omega - d\phi(\xi) = 0$. In other words,

$$d_G \tilde{\omega} = 0 \quad \Leftrightarrow \quad d\omega = 0 \quad \text{and} \quad \iota_\xi \omega = -d\phi^\xi.$$

In the language of symplectic geometry this says that ϕ is a **moment map** for the action of g and the closed form ω. If, in addition to being closed, the form ω is non-degenerate (and so symplectic) we say that the d_G closed form $\tilde{\omega}$ is an **equivariant symplectic form**. In other words, an equivariant symplectic form is a G-invariant symplectic form together with a moment map. Even if the form ω is not symplectic, we can call the ϕ occurring in $\tilde{\omega} = \omega - \phi$ the moment map.

9.2   The Case $M = G$

Suppose that $\tilde{\omega}$ is an equivariantly closed two-form on G (with G acting on itself by left multiplication). We know that $H^2_G(G) = H^2(G/G) = H^2(\text{pt}) = 0$, so every equivariantly closed two-form is equivariantly exact, i.e.

$$\tilde{\omega} = d_G\theta \tag{9.1}$$

for some equivariant one-form θ. But an equivariant one-form is just an invariant one-form. In other words the above equation holds with θ some left-invariant one-form. Now

$$d_G\theta = d\theta - \iota_{\xi_R}\theta$$

where ξ_R is the vector field corresponding to the left multiplication action of G on itself (and so is right invariant). In particular, the θ occurring in (9.1) is unique, since $d_G\theta = 0$ implies that $\iota_{\xi_R}\theta = 0$, and the $\{\xi_R\}$ span the tangent space at each point of G.

A left invariant one-form θ is determined by its value, $\ell = \theta(e) \in TG^*_e = g^*$ at the identity, and every $\ell \in g^*$ gives rise to a left invariant one form which we shall denote by θ_ℓ. Thus the most general equivariantly closed two-form on G is given by

$$d_G\theta_\ell = d\theta_\ell - \psi_\ell \quad \text{where} \quad \langle\psi_\ell, \xi\rangle = \iota_{\xi_R}(\theta_\ell). \tag{9.2}$$

Notice that
$$\psi_\ell(e) = \ell. \tag{9.3}$$

The G-equivariance of ψ_ℓ says that for any $b \in G$ and any $\xi \in g$,

$$[b\psi_\ell(\xi)](\cdot) := \psi_\ell(\xi)(b^{-1}\cdot) = \psi_\ell(\text{Ad}_b\xi)(\cdot).$$

Taking $b = a^{-1}$ and evaluating at e gives

$$\langle\psi_\ell(a), \xi\rangle = \langle\ell, \text{Ad}_{a^{-1}}\xi\rangle.$$

The coadjoint representation is defined as the contragredient of the adjoint representation and is given by

$$\langle\text{Ad}^\sharp_a\ell, \xi\rangle := \langle\ell, \text{Ad}_{a^{-1}}\xi\rangle.$$

So we have

$$\psi_\ell(a) = \mathrm{Ad}_a^\sharp \ell. \tag{9.4}$$

Let G_ℓ denote the stabilizer group of ℓ in the coadjoint representation. Then (9.4) implies that ψ_ℓ is constant on left cosets aG_ℓ. In other words, ψ_ℓ is in effect a map

$$\psi_\ell : \; G/G_\ell \to g^*. \tag{9.5}$$

We now prove a similar result for $d\theta_\ell$:

Theorem 9.2.1 *With the above notations*

1. *$d\theta_\ell$ is the pullback via the canonical projection of a closed two-form $\omega_\ell \in \Omega^2(G/G_\ell)$.*

2. *ω_ℓ is symplectic.*

Proof. Let G_ℓ act on G on the right, and consider G as a principal G_ℓ bundle:

$$\pi : G \to G/G_\ell.$$

To prove part 1) it suffices to prove that $d\theta_\ell \in \Omega^2(G)_{\mathrm{bas}}$, i.e to prove that $d\theta_\ell$ is G_ℓ-invariant and satisfies

$$\iota_{\xi_L} d\theta_\ell = 0 \tag{9.6}$$

for all $\xi \in g_\ell$, where, for any $\xi \in G$, we let ξ_L denote the left-invariant vector field corresponding to ξ which is determined by the right multiplication action of G on itself. Now since θ_ℓ is invariant under G_ℓ, so is $d\theta_\ell$. So it is more than enough to prove the stronger assertion

Lemma 9.2.1 *For any $\xi \in g$, (9.6) holds if and only if $\xi \in g_\ell$.*

To prove the lemma we use the identity

$$d\theta_\ell(\xi_L, \eta_L) = L_{\xi_L}[\theta_\ell(\eta_L)] - L_{\eta_L}[\theta_\ell(\xi_L)] - \theta_\ell([\xi_L, \eta_L]).$$

Since θ_ℓ, ξ_L and η_L are all left invariant, the first two terms on the right vanish and the third term is constant. Evaluating it at e give

$$d\theta_\ell(\xi_L, \eta_L) = -\langle \ell, [\xi, \eta] \rangle.$$

thus (9.6) holds if and only if

$$\left(\mathrm{ad}^\sharp(\xi)\ell \right)(\eta) = 0, \quad \forall \eta \in g$$

which says that $\xi \in g_\ell$. This proves the lemma, and with it statement 1) of the theorem.

So there is an $\omega_\ell \in \Omega(G/G_\ell)$ with $\pi^*\omega_\ell = d\theta_\ell$. We must prove that it is symplectic. For this it is enough to show that if $\pi^*\omega_\ell = d\theta_\ell$ is annihilated by a left invariant vector field, this vector field is tangent to the fibers of $G \to G/G_\ell$. But this is precisely the assertion of our lemma. $\square$.

Combining ω_ℓ with the map (9.5) gives an *equivariant* symplectic form

$$\tilde{\omega}_\ell := \omega_\ell - \psi_\ell \qquad (9.7)$$

on G/G_ℓ such that

$$\pi^* \tilde{\omega}_\ell = d_G \theta_\ell. \qquad (9.8)$$

We can state Theorem 9.2.1 in terms of coadjoint orbits, which was the original formulation by Kirillov, Kostant, and Souriau: Let $\mathcal{O}$ denote the coadjoint orbit containing ℓ. By (9.4) and (9.5), ψ_ℓ is a G-equivariant diffeomorphism of G/G_ℓ onto $\mathcal{O}$. Let

$$\varrho_\ell : \mathcal{O} \to G/G_\ell$$

denote the inverse diffeomorphism, so that

$$\psi_\ell \circ \varrho_\ell = i_{\mathcal{O}} : \mathcal{O} \to g^*$$

is the inclusion map of $\mathcal{O}$ into g^*. Let

$$\tilde{\omega} := \varrho_\ell^* \tilde{\omega}_\ell.$$

Then Theorem 9.2.1 becomes

Theorem 9.2.2 Kirillov-Kostant-Souriau *There is a unique equivariant symplectic form $\tilde{\omega}_{\mathcal{O}} \in \Omega_G^2(\mathcal{O})$ with moment map given by the inclusion, $i_{\mathcal{O}} : \mathcal{O} \to g^*$.*

Remark. Notice that in this section we did not need to assume that G was compact. There are, however, some special features of the orbit picture which are particularly nice when G is compact; see Section 9.4 below.

9.3   Equivariantly Closed Two-Forms on Homogeneous Spaces

Let K be a closed subgroup of G and suppose that $\tilde{\omega}$ is an equivariantly closed two-form on G/K. If $\pi : G \to G/K$ is the projection onto cosets, then $\pi^* \tilde{\omega}$ is a closed equivariant two-form on G and hence of the form $d_G \theta_\ell$. If we write

$$\tilde{\omega} = \omega - \phi$$

then $\pi^* \phi = \psi_\ell$ where $\psi_l(a) = \mathrm{Ad}_a^\sharp \ell$. In particular, at the identity coset, $K = eK$ we have $\phi(K) = \ell$ and, by the equivariance of $\phi : G/K \to g^*$ we see that K, the stabilizer group of the identity coset, must be contained in G_ℓ, the stabilizer group of ℓ. So we have the projection map

$$\rho : G/K \to G/G_\ell, \quad \rho(aK) = aG_\ell$$

and the commutative diagram

$$\begin{array}{ccc} & G & \\ {}^{\pi}\swarrow & & \searrow^{\pi_\ell} \\ G/K & \xrightarrow{\ \rho\ } & G/G_\ell. \end{array}$$

It follows that

$$\tilde{\omega} = \rho^* \tilde{\omega}_\ell = \rho^* \omega_\ell - \rho^* \phi_\ell.$$

If we identify G/G_ℓ with $\mathcal{O}_\ell$ we may think of ϕ_ℓ as the identity map, and hence the preceding equation implies that we may identify ρ with ϕ. We have proved:

Let $\tilde{\omega} = \omega - \phi$ be a closed equivariant two-form on G/K. Then ϕ is a G-equivariant map of G/K onto some coadjoint orbit $\mathcal{O} = G \cdot \ell = \mathcal{O}_\ell$ and

$$\tilde{\omega} = \phi^* \tilde{\omega}_\ell. \tag{9.9}$$

In particular, $\tilde{\omega}$ is symplectic if and only if the map ϕ is a covering map.

9.4   The Compact Case

In case G (and hence K) are compact, and G is connected, the preceding results can be strengthened. We will prove that every coadjoint orbit is simply connected which will imply that any cover (as in the last section) must be a bijection and that every G_ℓ is connected. We begin with a result which is of interest in its own right.

Let T be a maximal torus of the connected compact Lie group, G and let $\mathcal{O}$ be a coadjoint orbit. Then the action of T on $\mathcal{O}$ has a finite number of fixed points which are all non-degenerate, in the sense that the linear action of T on the tangent space at each fixed point has no non-zero fixed vectors.

Proof. The non-degeneracy assertion follows from the finiteness of the fixed points, since we may choose a T-invariant Riemann metric on $\mathcal{O}$ and then any line of fixed vectors in the tangent space is carried by the exponential map of this metric into a whole curve of fixed points. By a choice of an invariant scalar product on g we may identify the adjoint and coadjoint representation, and by the conjugacy theorem for Cartan subalgebras we may assume that $\mathcal{O} = \mathcal{O}_\ell$ with $T \subset G_\ell$. So T is a maximal torus of G_ℓ. A coset bG_ℓ is fixed by T if and only if $b^{-1}Tb \subset G_\ell$ and hence, by the conjugacy theorem for maximal compact subgroups of G_ℓ, there exists an $a \in G_\ell$ such that $a^{-1}b^{-1}Tba = T$. Thus $bG_\ell = baG_\ell$ and ba normalizes T. But $N(T)/T = W(T)$, the Weyl group, is finite. Hence there are finitely many fixed cosets, bG_ℓ. □

Let V be the tangent space at a fixed point. Then V decomposes under T into a direct sum of two dimensional subspaces. The action on each two dimensional subspace is rotation through angle $\alpha \cdot \theta$ where $\alpha = (\alpha_1, \ldots, \alpha_n)$ is a row vector with integer coordinates not all zero and $\theta = (\theta^1, \ldots, \theta^n)^\dagger$ is typical element of $T = (S^1)^n$. We may choose a $\xi = (\xi^1, \ldots, \xi^n)^\dagger \in t$, the Lie algebra of T, such that the one parameter group generated by ξ is dense in T and such that $\alpha(\xi) \neq 0$ for any α at any fixed point. The only zeros of the vector field on $\mathcal{O}$ corresponding to ξ are the fixed points of T, and these are all non-degenerate zeros of ξ. Then ϕ^ξ is a non-degenerate Morse function with critical points at the fixed points of T, and the index at each critical point is even. It follows that the Morse-Whitney stratification of $\mathcal{O}$ associated with ϕ^ξ consists of a single open cell W whose complement has codimension two. Hence every closed curve in $\mathcal{O}$ can be deformed to a curve in W and then contracted to a point. Hence $\mathcal{O}$ is simply connected.

It follows from the homotopy long exact sequence for the fibration $G \to G/G_\ell$ that G_ℓ is connected. We recall the argument in our special case. We consider G_ℓ as the fiber over the identity coset, eG_ℓ. Given any two points P and Q in G_ℓ we can connect them by a curve in G since G is assumed to be connected. This curve projects onto a closed curve γ in G/G_ℓ based at the identity coset. By the simple connectedness of G/G_ℓ we can find a homotopy of this curve to the trivial curve. In other words there is a map of the unit square, $\square$, into G/G_ℓ whose restriction to the bottom edge is γ and whose restriction to the top edge is the constant curve, taking the constant value G_ℓ. The lifting property for fibrations implies that we can lift this to a map of $\square$ into G whose restriction to the left hand side and top is identically P and whose restriction to the bottom is the curve from P to Q that we started with. The restriction to the right-hand side then gives a curve joining P to Q in G_ℓ.

9.5   Minimal Coupling

Let M be a G-manifold on which G acts freely with quotient space $X = M/G$ and with projection

$$\pi : M \to X$$

sending each $m \in M$ into its G-orbit. We think of this as a principal G-bundle.

Let F be another G-manifold. The diagonal action of G on $M \times F$ is free, so we can form the quotient

$$W := (M \times F)/G$$

and mapping

$$\gamma : W \to X, \qquad \gamma([(m, f)]) = \pi(m) \tag{9.10}$$

where $[(m, f)]$ denotes the equivalence class of (m, f). This makes W into a fiber bundle over X with typical fiber F. In the language of the topologists,

W is a "twisted product" of X and F. If X and F are symplectic manifolds, then there is a product symplectic structure on $X \times F$. In this section we shall prove an analogous result for the twisted product, under the assumption that we have a symplectic form μ on X and an *equivariant* symplectic form $\tilde{\omega}$ on F. Let pr_1 and pr_2 denote the projections of $M \times F$ onto the first and second factors. Then for each $\epsilon \in \mathbf{R}$ we get the equivariant two-form

$$\mu_\epsilon := (\mathrm{pr}_1)^* \pi^* \mu + \epsilon (\mathrm{pr}_2)^* \tilde{\omega} \tag{9.11}$$

on $M \times F$.

Let us equip M with a connection, and let $\theta^1, \ldots, \theta^n$ be the connection forms (relative to a choice of basis of g). The connection on M induces a connection on the bundle

$$M \times F \to W$$

whose connection forms are

$$(\mathrm{pr}_1)^* \theta^1, \ldots, (\mathrm{pr}_1)^* \theta^n.$$

As we proved in Chapter 5, this connection allows us to define a Cartan map

$$\Omega_G(M \times F) \to \Omega(W). \tag{9.12}$$

Applying this map to μ_ϵ gives us a closed two-form

$$\nu_\epsilon \in \Omega^2(W)$$

which is called the **minimal coupling form** with coupling constant ϵ.

Theorem 9.5.1 *For $|\epsilon| \neq 0$ sufficiently small, ν_ϵ is symplectic.*

Proof. Let $\omega \in \Omega^2(W)$ be the form obtained by applying the Cartan map (9.12) to $(\mathrm{pr}_2)^* \tilde{\omega}$. Then if

$$2d := 2p + 2q = \dim W, \quad 2p := \dim X, \quad 2q := \dim F$$

we have

$$\nu_\epsilon^d = \epsilon^q \omega^q \wedge (\gamma^* \mu)^p + O(\epsilon^{q+1})$$

and it is easy to see that the first term on the right is nowhere zero. □

For applications of this minimal coupling form to elementary particle physics see [St1] or [GS]. For applications to representation theory, see [GLS]. For applications to topology see [GLSW].

9.6 Symplectic Reduction

Let M be a G-manifold on which G acts freely and let $\tilde{\omega} = \omega - \phi$ be an equivariant symplectic form with moment map $\phi : M \to g^*$. We will show that ϕ is a submersion, i.e. that

$$d\phi_p : TM_p \to Tg^*_{\phi(p)} \cong g^*$$

is surjective. Indeed, if not, the components of ϕ with respect to a basis $\xi_1, \ldots, \xi_n$ of g would have to be functionally dependent at p. Put another way, some non-trivial linear combination of

$$(d\phi^{\xi_1})_p, \ldots, (d\phi^{\xi_n})_p$$

must vanish, i.e.

$$(d\phi^\xi)_p = 0$$

for some $\xi \neq 0 \in g$. But

$$d\phi^\xi = -\iota_{\xi_M}\omega$$

where ξ_M is the vector field on M corresponding to ξ. Since ω is symplectic, we conclude that $(\xi_M)_p = 0$, contradicting the assertion that the action of G is free. $\square$

Since ϕ is G-equivariant the level set

$$M_0 := \phi^{-1}(0)$$

is a G-invariant submanifold of M. Let

$$X_0 := M_0/G.$$

We have the inclusion

$$i_0 : M_0 \to M$$

and the projection

$$\pi_0 : M_0 \to X_0.$$

Theorem 9.6.1 Marsden-Weinstein   *There exists a unique symplectic form, ν_0 on X_0 with the property*

$$\pi_0^* \nu_0 = i_0^* \omega. \tag{9.13}$$

Proof. The equivariant form $i_0^* \tilde{\omega} \in \Omega_G(M_0)$ is d_G-closed. Since $i_0^* \phi = 0$ by the definition of M_0, we have

$$i_0^* \tilde{\omega} = i_0^* \omega.$$

Since $d\omega = 0$ we have

$$0 = d_G i_0^* \omega = \iota_a(i_0^* \omega) \otimes x^a,$$

where $x^1, \ldots, x^n$ is the dual basis to our basis $\xi_1, \ldots, \xi_n$ of g. Thus

$$\iota_a(i_0^* \omega) = 0 \quad \forall a = 1, \ldots, n,$$

in other words $i_0^* \omega$ is basic with respect to the fibration

$$M_0 \to X_0.$$

This implies that there exists a two-form ν_0 on X_0 satisfying (9.13). To prove that ν_0 is symplectic we use the identity

$$\iota_1 \cdots \iota_n (\omega)^d = (d!/n!)\omega^{d-n} \wedge d\phi^1 \wedge \cdots \wedge d\phi^n \qquad \text{at all } m \in M_0 \qquad (9.14)$$

where $\dim M = 2d$. (We leave the proof of this identity as an easy exercise.) In fact, $\dim X_0 = \dim M - n - n = 2(d-n)$, so to prove that ν_0 is symplectic, it suffices to show that ν_0^{d-n} is nowhere vanishing, which is the same as proving that $(\pi_0^* \nu_0)^{d-n}$ is nowhere vanishing. Since $\pi_0^* \nu_0 = i_0^* \omega$, this is the same as proving that $(i_0^* \omega)^{d-n}$ vanishes nowhere on M_0. This is the same as showing that the right hand side of (9.14) does not vanish at any point of M_0. But the left hand side of (9.14) vanishes nowhere on M, since the ξ_a are everywhere independent and ω is symplectic.   □

The operation of passing from

$$(M, \tilde{\omega}) \quad \text{to} \quad (X_0, \nu_0)$$

is known as **symplectic reduction** or Marsden-Weinstein reduction. For a more detailed treatment see [GS] section 26.

If the group G is abelian, there is nothing sacrosanct about the *zero* level set of the moment map. For every $a \in g^*$ the submanifold

$$M_a := \phi^{-1}(a) \qquad (9.15)$$

is G-invariant. Let

$$X_a := M_a/G \qquad (9.16)$$

and let

$$i_a : M_a \to M, \qquad \pi_a : M_a \to X_a$$

the inclusions and projections. Then the proof of Theorem 9.6.1 goes over unchanged to prove

Theorem 9.6.2 Marsden-Weinstein *There exists a unique symplectic form, ν_a on X_a with the property*

$$\pi_a^* \nu_a = i_a^* \omega. \qquad (9.17)$$

If the moment map ϕ is proper, the level set M_a is compact, and hence so is its quotient space X_a. In particular, its symplectic volume

$$\frac{1}{(d-n)!} \int_{X_a} \nu_a^{d-n} \qquad (9.18)$$

is well defined and is a smooth function of a. In the next section we will prove the following theorem of Duistermaat and Heckman:

Theorem 9.6.3 *The symplectic volume (9.18) is a polynomial as a function of a.*

9.7   The Duistermaat-Heckman Theorem

Let
$$X := M/G$$
and
$$\pi : M \to X$$
be the canonical projection. If G is abelian, the coadjoint action is trivial. Hence the fact that the moment map $\phi : M \to g^*$ is G-equivariant means that it is invariant. Thus it factors through π, i.e. there is a smooth map
$$\psi : X \to g^*$$
such that
$$\phi = \psi \circ \pi. \tag{9.19}$$
Since ϕ is a submersion, so is ψ and hence, by (9.19),
$$M_a = \phi^{-1}(a) = \pi^{-1}\left(\psi^{-1}(a)\right)$$
so
$$X_a = \psi^{-1}(a). \tag{9.20}$$
Let
$$j_a : X_a \to X$$
denote the inclusion, so that we have the commutative diagram

$$
\begin{array}{ccc}
M_a & \xrightarrow{\ i_a\ } & M \\
\pi_a \downarrow & & \downarrow \pi \\
X_a & \xrightarrow[\ j_a\ ]{} & X
\end{array}
\tag{9.21}
$$

Now let us equip M with a connection and consider the associated Cartan map
$$\Omega_G(M) \to \Omega(X). \tag{9.22}$$
Under this map, the equivariant symplectic form
$$\tilde{\omega} = \omega - \phi_r x^r$$
gets mapped to
$$\nu - \psi_r \mu^r \tag{9.23}$$
where the μ^r are the curvature forms of the connection, and where ν is the unique form on X with the property that
$$\pi^* \nu = \omega_{\text{hor}}. \tag{9.24}$$

From (9.17), (9.21), and (9.24) we obtain

$$j_a^* \nu = \nu_a. \tag{9.25}$$

Let c denote the cohomology class of $\nu - \psi_r \mu^r$, and let $[\mu^i]$ denote the cohomology class of the curvature form μ^i. Letting $[\nu_a]$ denote the cohomology class of ν_a, we conclude that

$$[\nu_a] = j_a^* \left(c + a_r [\mu^r] \right). \tag{9.26}$$

In other words, $[\nu_a]$ "varies linearly with a".

Since X_a is compact and oriented, the embedding $j_a : X_a \to X$ defines a *homology* class

$$[X_a] \in H_{2(d-n)}(X, \mathbf{Z})$$

in the integer homology group of dimension $2(d - n)$. This homology class depends smoothly on a, and being an integer class is thus independent of a. So let us fix an a_o in the image of ϕ. Then

$$[X_a] = [X_{a_0}]. \tag{9.27}$$

We can now prove the Duistermaat-Heckman theorem. The integral (9.18) can be interpreted topologically as the pairing of the constant homology class $[X_{a_0}]$ with the cohomology class

$$\frac{1}{(d-n)!} \left(c + a_r [\mu^r] \right)^{d-n}$$

so the value is clearly a polynomial of degree $d - n$.

9.8 The Cohomology Ring of Reduced Spaces

Let (M, ω) be a symplectic manifold of dimension $2d$, G a compact connected Lie group, $\tau : G \to \mathrm{Diff}(M, \omega)$ a Hamiltonian action of G on M with moment map $\phi : M \to g^*$ and $Z = \phi^{-1}(0)$ the zero level set of the moment map. If 0 is a regular value of ϕ, then the action of G on Z is locally free, and the reduced space

$$X = Z/G$$

is a symplectic orbifold of dimension $2(d-n)$ where $n = \dim G$. Let us assume that ϕ is proper. Then X is compact, so its cohomology ring, $H^*(X, \mathbf{C})$ is finite dimensional and satisfies Poincaré duality. In the early 1980's, Kirwan, [Ki], showed how to compute the Betti numbers of X, using Morse theoretic techniques. Recently, quite a bit of progress has been made on the much more difficult problem of understanding the cohomology ring structure of X. Some relevant papers are [Wi], [Ka], [TW], and [JK]. See also the survey paper, [D]. In particular, Jeffrey and Kirwan have found a general formula for pairings of cohomology classes on X which enables them, in principle, to determine the ring structure of X when the ambient space, M is compact. (In practice, the problem of decoding the ring structure from their formula is non-trivial.)

The purpose of this section is to point out that a good deal of information about the cohomology ring structure of X can be extracted from the Duistermaat-Heckmann theorem. The version of this theorem that we will use is a slight sharpening of the version we proved in the last section: Let $G = T^n$ be the standard n-dimensional torus, and for $\ell \in g^*$ close to zero, set

$$Z_\ell = \phi^{-1}(\ell) \quad \text{and} \quad X_\ell = Z_\ell/G. \tag{9.28}$$

Theorem 9.8.1 *As a differentiable manifold $X =: X_0 = X_\ell$ and*

$$[\mu_\ell] = [\mu] + \sum \ell_i c_i \tag{9.29}$$

where $[\mu_\ell]$ is the cohomology class of the symplectic form on X_ℓ, $[\mu] = [\mu_0]$ is the cohomology class of symplectic form on X, and $c = (c_1, \ldots, c_n)$ is the Chern class of the fibration $Z \to X$.

Let

$$v(\ell) = \int_{X_\ell} \exp([\mu_\ell]) = \int_X \exp([\mu] + \sum \ell_i c_i) \tag{9.30}$$

be the symplectic volume of X_ℓ. From (9.30) it follows that $v(\ell)$ is a polynomial of degree $d - n$ and that

$$\left(\frac{\partial^\alpha}{\partial \ell^\alpha} v\right)(0) = \frac{1}{k!} \int_X [\mu]^k \cdot c_1^{\alpha_1} \cdots c_n^{\alpha_n},$$
$$|\alpha| = \alpha_1 + \cdots + \alpha_n = d - n - k, \quad 0 \le k \le d - n. \tag{9.31}$$

In particular, if $|\alpha| = d - n$

$$\left(\frac{\partial^\alpha}{\partial \ell^\alpha} v\right)(0) = \int_X c_1^{\alpha_1} \cdots c_n^{\alpha_n} \tag{9.32}$$

Thus the coefficients of degree $d - n - k$ of the polynomial $v(\ell)$ determine the cohomology pairings (9.31) and, in particular, the leading coefficients determine the cohomology pairings (9.32).

The identities (9.31) and (9.32) are more or less well known, but they haven't been used very much as a tool for computing the ring structure of $H^*(X, \mathbf{C})$. Possibly this is because they give no information if the fibration $Z \to X$ is trivial, which is frequently the case. On the other hand, if the c_i generate the cohomology ring, one can read off from (9.32) all multiplicative relations of the form

$$\sum_{|\beta| + |\gamma| = d - n} a_\beta c^{\beta + \gamma} = 0, \quad |\beta| + |\gamma| = d - n,$$

and hence, by Poincaré duality, all relations of the form

$$\sum a_\beta c^\beta = 0, \quad 0 \le |\beta| \le d - n.$$

Specifically,

$$\int \sum_{|\beta|+|\gamma|=d-n} a_\beta c^{\beta+\gamma} = \left(\frac{\partial}{\partial \ell^\gamma} \sum a_\beta \frac{\partial^\beta v}{\partial \ell^\beta} \right)(0)$$

$$= \left(\frac{\partial}{\partial \ell^\gamma} Q v_{top} \right)(0)$$

where v_{top} denotes the top order homogeneous part of v and where $Q = \sum a_\beta \frac{\partial}{\partial \ell^\beta}$. Clearly this last expression vanishes for all $|\gamma| = d - n - |\beta|$ if and only if $Q v_{top} = 0$. From this one obtains:

Theorem 9.8.2 *If $c_1, \ldots, c_n$ generate $H^*(X, \mathbf{C})$, then $H^*(X, \mathbf{C})$ is isomorphic as an abstract ring to*

$$\mathbf{C}[x_1, \ldots, x_n]/\mathrm{ann}(v_{top}) \tag{9.33}$$

where $Q(x_1, \ldots, x_n) \in \mathrm{ann}(v_{top})$ if and only if $Q(\frac{\partial}{\partial \ell_1}, \ldots, \frac{\partial}{\partial \ell_n})v_{top}(\ell) = 0$.

In this section we discuss two examples of applications of this theorem in the first of which X is the generic flag variety and in the second X is the toric variety associated with a simplicial fan. The cohomology rings of the flag manifolds were determined by Borel in his classic paper, [Bo], and for toric varieties by Danilov in [Dan]. It is interesting to note that the structure of $H^*(X, \mathbf{C})$ in both these examples is given by the general recipe (6).

Let us return to the case of an arbitrary (M, ω). Let p be a fixed point of the $G = T^n$ action on M, and let $\alpha_1, \ldots, \alpha_d$ be the weights of the isotropy representation of G on the tangent space at p. We will call p an *extremal fixed point* if there exists a $v \in g$ such that

$$\alpha_i(v) > 0, \quad \forall i.$$

(See [GLS], Section 3.) If p has this property, we will prove in Section 9.8.3 that if ξ is a regular value of ϕ close to $\phi(p)$ then X_ξ is a toric variety, and we will show how to compute its moment polytope in terms of the α_i. This has the following useful corollary. Suppose that M is compact and the fixed point set of G is finite. By the convexity theorem ([At] or [GS]) the image, $\Delta = \phi(M)$, is a convex polytope. Let $\Delta^0 \subset \Delta$ denote the set of regular values of ϕ. The connected components, $\Delta_1, \ldots, \Delta_N$ of Δ^0 are themselves open convex polytopes. By the Duistermaat-Heckmann theorem, the diffeotype of the reduced space, X_ξ is constant as ξ varies over each Δ_i. In particular, the cohomology ring of this reduced space depends only on Δ_i.

Theorem 9.8.3 *Suppose the closure of Δ_i contains a vertex of Δ. Then its associated reduced space is toric variety.*

Thus for a Δ_i whose closure contains a vertex, we can compute the cohomology ring of its reduced space by theorem 9.8.6 below. In general, there will be many connected components of Δ^0 which *don't* have this property. This brings up the following interesting question: How does the cohomology ring of the reduced space, X_ξ, change, as ξ passes through a common $(n-1)$-dimensional face of two adjacent Δ_i's? For some recent results concerning this question see [TW].

9.8.1   Flag Manifolds

Let K be a compact semi-simple Lie group and T its Cartan subgroup. The adjoint action of T on the lie algebra k has a T-equivariant splitting

$$k = t \oplus t^\perp. \tag{9.34}$$

Here t is the Lie algebra of T and $t^\perp$ is its orthocomplement with respect to the Killing form. Using the right action of T on K we get the fibration

$$K \to K/T \tag{9.35}$$

mapping an element of K into its right T coset. This gives us a fibration of K over the flag variety K/T. The splitting (9.34) gives us a splitting of the tangent space to K at e into vertical and horizontal components, which can be extended via the left action of K to the whole of K, giving an intrinsic $K \times T$ invariant connection, θ, on the bundle (9.35). By minimal coupling, cf. Section 9.5, one gets a presymplectic form

$$d\langle \theta, \mathrm{pr}_2 \rangle \tag{9.36}$$

on $K \times t^*$ where

$$\mathrm{pr}_2 : K \times t^* \to t^* \tag{9.37}$$

denotes projection onto the second factor. The presymplectic form given by (9.36) is symplectic on the set

$$K \times (t_+^*)^0, \tag{9.38}$$

where t_+^* denotes the positive Weyl chamber and t_+^0 its interior. Letting $K \times T$ act trivially on the second factor in (9.38) gives a Hamiltonian action of $K \times T$ on $K \times (t_+^*)^0$. The T-moment map for this action is (the restriction of) pr_2.

For $\ell \in (t_+^*)^0$ let O_ℓ denote the coadjoint orbit of K through ℓ. The stabilizer of ℓ in K is T, so

$$O_\ell \sim K/T$$

as a K-homogeneous space. The reduction of $K \times (t_+^*)^0$ at ℓ (with respect to T) is also K/T as a homogeneous K space. We claim

Proposition 9.8.1 *The reduced space of $K \times (t_+^*)^0$ with respect to T at ℓ is isomorphic as a Hamiltonian K space to O_ℓ.*

Proof. At the point $(e, \ell) \in K \times (t_+^*)^0$ the symplectic form defined by minimal coupling satisfies

$$\omega(x, y) = -\langle \ell, [x, y] \rangle, \quad x, y \in k/t. \tag{9.39}$$

But the expression on the right is precisely the Kirillov-Kostant form evaluated on $x, y \in k/t$ at $\ell \in O_\ell$. Let c be the Chern class of the T fibration $K \to K/T$. By definition, this is a t-valued cohomology class of degree two which we can write as

$$c = \sum H_i \otimes c_i, \quad c_i \in H^2(K/T, \mathbf{R}),$$

where H_i is the standard Weyl basis of t.

Proposition 9.8.2 $c_1, \ldots, c_n$ *generate the cohomology ring* $H^*(K/T, \mathbf{C})$.

For a proof of this see Borel [Bo2].

As a corollary, we see that the identities (9.32) completely determine the cohomology ring structure of K/T. To make these identities more explicit, we will use the following result, cf. [BGV] p.232, giving a formula for the symplectic volume of a coadjoint orbit, and hence an explicit formula for the left hand sides of (9.31) and (9.32):

Proposition 9.8.3 *The symplectic volume, $v(\ell)$ of the coadjoint orbit, O_ℓ is given by the formula*

$$v(\ell) = \frac{1}{\gamma} \prod_{\alpha \in \Pi^+} (\ell, \alpha) \tag{9.40}$$

where α ranges over the positive roots, where (ℓ, α) is the inner product of ℓ and α with respect to the form induced on t^* by the negative of the Killing form and where the constant γ is given by

$$\gamma = \prod_{\alpha \in \Pi^+} (\rho, \alpha), \quad \rho = \frac{1}{2} \sum_{\alpha \in \Pi^+} \alpha. \tag{9.41}$$

Notice that the right hand side of (9.40) is a polynomial in ℓ of degree $d - n$ as required, as $2d = \dim K + n$, $\dim K = 2p + n$ where p denotes the number of positive roots. Plugging (9.40) into (9.32) gives

$$\int c_1^{\beta_1} \cdots c_n^{\beta_n} = \frac{1}{\gamma} \frac{\partial^{|\beta|}}{\partial \ell^\beta} \prod_{\alpha \in \Pi^+} (\ell, \alpha). \tag{9.42}$$

We will say a few words about how these results are related to the theorem of Borel which we mentioned above. Borel's theorem says that the cohomology ring of K/T is isomorphic to the ring

$$\mathbf{C}[x_1, \ldots, x_n] / \mathbf{C}[x_1, \ldots, x_n]^W, \tag{9.43}$$

$\mathbf{C}[x_1, \ldots, x_n]^W$ being the ideal generated by the Weyl group invariant polynomials of degree greater than zero. To deduce this from Theorem 1.2, one has to show that if $Q(x_1, \ldots, x_n)$ is a homogeneous Weyl group invariant polynomial of degree greater than zero and $v(\ell)$ is the function (12) then

$$Q\left(\frac{\partial}{\partial \ell}\right) v(\ell) = 0 .$$

Notice, however, that if α is a simple root and $\sigma_\alpha : t \longrightarrow t$ is reflection through the hyperplane $\alpha(\ell) = 0$, then σ_α maps all of the positive roots except α into themselves and maps α onto $-\alpha$. Hence

$$\sigma_\alpha^* v(\ell) = -v(\ell) ,$$

and since Q is Weyl group invariant,

$$\sigma_\alpha^* Q\left(\frac{\partial}{\partial \ell}\right) v(\ell) = -v(\ell) .$$

In particular, $Q\left(\frac{\partial}{\partial \ell}\right) v(\ell)$ vanishes on the hyperplane $\alpha(\ell) = 0$, so the monomial $\alpha(\ell)$ divides $Q\left(\frac{\partial}{\partial \ell}\right) v(\ell)$. Since every positive root is Weyl group conjugate to a simple root, it follows that, for every root α, $\alpha(\ell)$ divides $Q\left(\frac{\partial}{\partial \ell}\right) v(\ell)$. Hence $v(\ell)$ itself divides $Q\left(\frac{\partial}{\partial \ell}\right) v(\ell)$, and this is impossible unless $Q\left(\frac{\partial}{\partial \ell}\right) v(\ell) = 0$.

Remark: For this argument we are indebted to David Vogan.

9.8.2   Delzant Spaces

We will refer to the toric varieties in this section as **Delzant spaces** since we will be thinking of them as symplectic manifolds (or orbifolds) rather than as complex projective varieties. Let us briefly review the definition of these spaces as given by Delzant [Del]. Let V be a real n-dimensional vector space and $L \subset V$ and n-dimensional lattice, with dual space V^* and dual lattice L^*. Let T be the torus $T = V/L$. Given a convex polytope, $\Delta \subset V^*$ satisfying certain axioms listed below, we will associate a symplectic orbifold, X_Δ of dimension $2n$ to Δ. We will equip X_Δ with a Hamiltonian T action so that Δ is the image its moment map. By the uniqueness theorem of Delzant this characterizes X_Δ up to isomorphism.

The assumptions about Δ are the following:

1. Exactly n edges of Δ meet at every vertex.

2. The edges meeting at any vertex p point in rational directions (relative to L^*), i.e. lie along half rays of the form

$$p + t\beta_i, \quad t \geq 0, \quad \beta_i \in L^*, \quad i = 1, \ldots, n.$$

3. The $\beta_i(p)$, $i = 1, \ldots, n$, form a a basis of V^*.

If one wants X_Δ to be a manifold rather than merely an orbifold, condition **3** should be replaced by the stronger

3*. The $\beta_i(p)$, $i = 1, \ldots, n$, form a basis of L^*.

Let d denote the number of $(n - 1)$-dimensional faces of Δ. Condition **2** implies that these faces can be defined by equations of the form

$$\langle \ell, u_i \rangle = \lambda_i \quad i = 1, \ldots, d,$$

where the vector u_i belongs to L. The vector u_i can be normalized by requiring it to be a primitive element of L, which then determines it up to sign. The sign can be fixed by requiring that Δ be contained in the half space

$$\langle \ell, u_i \rangle \geq \lambda_i, \quad i = 1, \ldots, d. \tag{9.44}$$

These normalizations determine the u_i and λ_i uniquely. Conversely, Δ is given as the intersection of the half spaces (9.44).

Define the map

$$\pi : \mathbf{R}^d \to V, \quad \pi : e_i \mapsto u_i \tag{9.45}$$

where $\{e_1, \ldots, e_d\}$ is the standard basis of $\mathbf{R}^d$. Since

$$\pi : \mathbf{Z}^d \to L,$$

we get an induced map

$$\Pi : T^d \to T, \quad T^d = \mathbf{R}^d / \mathbf{Z}^d. \tag{9.46}$$

We have the standard linear action, ρ, of the torus t^d on $\mathbf{C}^d$ given by

$$\rho(x)z = (e^{2\pi i x_1} z_1, \ldots, e^{2\pi i x_d} z_d), \tag{9.47}$$

where

$$x = (x_1, \ldots, x_d) \in T^d \quad \text{and} \quad z = (z_1, \ldots, z_d) \in \mathbf{C}^d.$$

This action preserves the symplectic form

$$-i \sum_j dz_j \wedge d\bar{z}_j$$

and is a Hamiltonian action with moment map

$$J(z) = \frac{1}{2} \sum |z_j|^2 e_j^*. \tag{9.48}$$

Let G be the kernel of the map Π (cf. (9.46)). Restricting ρ to G gives a Hamiltonian action of G on $\mathbf{C}^d$ with moment map

$$\phi(z) = \frac{1}{2} \sum |z_i|^2 \alpha_i \tag{9.49}$$

where each α_i is the weight of G on the one-dimensional subspace of $\mathbf{C}^d$ spanned by e_i, $i = 1, \ldots, d$. Set

$$\lambda = \sum \lambda_i \alpha_i$$

where the λ_i are the same as those entering into the description (9.44) of Δ. Let Z_λ be the λ level set of the map ϕ

$$Z_\lambda = \{z : \sum |z_i|^2 \alpha_i = \lambda\}.$$

We refer to [Del] or [Gu1] for the following:

Proposition 9.8.4 *If Δ satisfies conditions* **1-3** *the action of G on Z_λ is locally free. If, in addition,* **3*** *holds, then this action is free.*

We now define

$$X_\Delta = Z_\lambda / G. \tag{9.50}$$

By the proposition, this is an orbifold, and is a manifold if **3*** holds. Since G is the kernel of Π, the group T acts on X_Δ in Hamiltonian fashion. We refer again to [Del] or [Gu1] for a proof of

Proposition 9.8.5 *The image of the moment map for the T action on X_Δ is the polytope Δ.*

The aim of this subsection is to give a description of the cohomology ring of X_Δ along the lines indicated in the introduction to this section. To this end we need to determine the Chern class of the fibration

$$Z_\lambda \to X_\Delta.$$

Since Z_λ is a principal G-bundle, a one-dimensional representation of G gives rise to a line bundle over X_Δ. Let L_i be the line bundle associated to the representation of G on the one-dimensional subspace of $\mathbf{C}^d$ spanned by e_i (with weight α_i). Let c_i be the corresponding Chern class.

Proposition 9.8.6 *The $\mathbf{R}^d$-valued cohomology class*

$$c = (c_1, \ldots, c_d)$$

takes values in g, the Lie algebra of G, and is the Chern class of the fibration $Z_\lambda \to X_\Delta$.

Proof. We content ourselves with proving the first of these two assertions. We must show that if $\kappa = (\kappa_1, \ldots, \kappa_d) \in (\mathbf{R}^d)^*$ lies in the annihilator space, g^0 of g then $\sum \kappa_i c_i = 0$. It is enough to prove this when the $\kappa_1, \ldots, \kappa_d$ are integers. But then $\sum \kappa_i c_i$ is the Chern class of the line bundle

$$L_1^{\kappa_1} \otimes \cdots \otimes L_d^{\kappa_d}$$

which is the line bundle associated with the weight $\sum \kappa_i c_i$. But

$$\kappa \in g^0 \iff \sum \kappa_i \alpha_i = 0,$$

so this line bundle is the trivial bundle and hence its Chern class is zero. .

For the following see [Dan] section 11:

Proposition 9.8.7 *The Chern classes c_i generate the cohomology ring $H^*(X_\Delta, \mathbf{C})$.*

If we now let λ vary slightly about a fixed value, the symplectic volume varies, and we can apply (9.32). Once again, to apply this formula we need an alternative computation of the symplectic volume: Here we have a strikingly simple description, for whose proof we refer to [Gu1] section 6:

Theorem 9.8.4 *The symplectic volume of X_Δ is the Euclidean volume of the polytope Δ.*

So if we let $v(\lambda)$ denote this volume, we can write (9.32) as

$$\frac{\partial^\beta v(\lambda)}{\partial \lambda^\beta} = \int_{X_\Delta} c_1^{\beta_1} \cdots c_d^{\beta_d} \tag{9.51}$$

for any multi-index β with $|\beta| = n$. (Note that since $v(\lambda)$ is a polynomial of degree n the left hand side of (9.51) is a constant.)

We will briefly describe the tie-in between this result and the result of Danilov which we alluded to in the introduction of this section. Let g be the Lie algebra of G and ι the inclusion of g into $\mathbf{R}^n$. From (9.46) one gets an exact sequence

$$0 \leftarrow g^* \overset{\iota^*}{\leftarrow} \left(\mathbf{R}^d\right)^* \overset{\pi^*}{\leftarrow} V^* \leftarrow 0 \ .$$

For λ and λ' in $\mathbf{R}^d$ let Δ and Δ' be the polytopes defined by

$$\langle \ell, u_i \rangle \geq \lambda_i \quad \text{and} \quad \langle \ell, u_i \rangle \geq \lambda_i' \ .$$

We leave as an exercise the following result:

Proposition 9.8.8 *Δ and Δ' are congruent if and only if $\lambda - \lambda' \in \ker \iota^*$ in which case*

$$\Delta' = \Delta + v$$

where $\lambda - \lambda' = \pi^* v$.

From Proposition 9.8.8 one gets n first order partial differential equations satisfied by $v(\lambda)$ namely

$$D_{w_i} v(\lambda) = 0 \ , \quad i = 1, \ldots, n \ , \tag{9.52}$$

where $w_i = \pi^* v_i$ and v_i, $i = 1, \ldots, n$, is a basis of V; and from these equations one gets n homogeneous generators of degree two of the ideal, $\text{ann}(v)$. However, in addition to these generators, there are also some generators of higher order. Namely by (9.44) one can associate with each of the standard basis vectors, e_i, of $\mathbf{R}^d$ the $(n-1)$-dimensional face of Δ defined by

$$\langle \ell, u_i \rangle = \lambda_i \ .$$

Denoting this face by F_i it is not hard to show that for a multi-index, $I = (i_1, \ldots, i_k)$, $1 \le i_1 < i_2 < \ldots < i_k \le d$,

$$\frac{\partial}{\partial \lambda_I} v(\lambda) = 0 \qquad (9.53)$$

if and only if $F_{i_1} \cap \ldots \cap F_{i_k} = \emptyset$. We claim (but will not attempt to prove here):

Theorem 9.8.5 $\text{ann}(v_{top})$ *is generated by the n generators of degree two associated with the n equations (9.52) and the generators of degree $2|I|$ associated with the equations (9.53).*

Combining this with Theorem 9.8.2 one ends up with the theorem of Danilov:

Theorem 9.8.6 $H^*(X_\Delta, \mathbf{C})$ *is a ring with d generators (each of degree 2) satisfying the relations (9.52) and (9.53).*

9.8.3   Reduction: The Linear Case

Let V be a complex d-dimensional vector space equipped with a positive definite Hermitian form, H and let ω be the symplectic form $\omega = \text{Im}H$. Let G be an r-dimensional torus and ρ a unitary representation of G on V. We can ignore the linear structure on V and consider ρ as a Hamiltonian action. In this section we discuss what reduced spaces look like in this linear setting. Let $e_1, \ldots, e_d$ be an orthonormal basis of V with the property that the one-dimensional subspaces spanned by the e_i are G-invariant. Each such subspace has an associated weight, call it α_i. This basis gives us a coordinate system $(z_1, \ldots, z_d)$ on V.

Proposition 9.8.9 *The moment map for the action ρ is*

$$\phi : \ z \mapsto \sum |z_i|^2 \alpha_i. \qquad (9.54)$$

We will henceforth assume that the action of G is effective. (This implies that the weights α_i span g^*.) For the following see [GLS] section 3:

Proposition 9.8.10 *The moment map (9.54) is proper if and only if there exists a vector $v \in g$ for which $\alpha_i(v) > 0$, $\forall i$.*

The action ρ is said to be *quasifree* if, for all $p \in V$ the stabilizer group, G_p is either of dimension greater than zero or is trivial. It turns out that for linear actions quasifree is equivalent to the condition that G_p be connected for every $p \in G$; cf. [GPS]. We also refer to [GPS] for the following

Proposition 9.8.11 ρ *is quasi-free if and only if every r-element subset of* $\{\alpha_1, \ldots, \alpha_d\}$ *which spans g^* is also a set of generators for the weight lattice of G.*

It is also clear from (9.54) that the image of ϕ is the cone

$$\{x_1\alpha_1 + \cdots + x_d\alpha_d; \quad x_i \geq 0\}. \tag{9.55}$$

For the following more refined result see [GLS], Proposition 3.28,

Proposition 9.8.12 *If $\alpha_{i_1}, \ldots, \alpha_{i_m}$ span a proper subspace of g^* every point on the cone*

$$\{\sum y_q \alpha_{i_q}; \quad y_q \geq 0\}$$

is a critical value of the moment map (9.54). Conversely every critical value lies on one of these cones.

Now let λ be a regular value of the moment map. If the moment map is proper, the level set

$$Z_\lambda = \phi^{-1}(\lambda)$$

is a compact submanifold of V. If the action is also quasi-free, the action of G on Z_λ is free, and the reduced space

$$X_\lambda = Z_\lambda / G \tag{9.56}$$

is a compact symplectic manifold. We will prove that this manifold is a Delzant space (of the type considered in the preceding section): Let $\mathbf{R}_+^d$ denote the positive orthant

$$\{(x_1, \ldots, x_d), \quad x_i \geq 0\}$$

and let

$$\tau : \mathbf{R}^d \to g^*$$

be the linear map sending the ith standard basis element into α_i. Let

$$\Delta_\lambda = \tau^{-1}(\lambda) \cap \mathbf{R}_+^d.$$

Theorem 9.8.7 X_λ *is a Delzant space and its moment polytope is Δ_λ.*

Proof. In terms of our coordinates, the representation given by (9.47) is a representation of T^d on V and we can regard G as a subgroup of T^d. The representation given by (9.47) restricts to the given representation of G, so there is no harm in denoting this extended representation also by ρ.

This action of T^d commutes with the action of G and induces a Hamiltonian action of T^d/G on X_λ. Now T^d acts freely near any point $z \in V$ if all its coordinates $z_i \neq 0$, and hence T^d acts freely on an open dense subset of Z_λ. Consequently T^d/G act freely on an open dense subset of X_λ. Since $2n = \dim X_\lambda = 2\dim T^d/G$ where $n = d - r$ the action of T^d/G is a Delzant action, cf. [Gu1]. The computation of its moment polytope, which we will omit, involves staring carefully at the Delzant construction which we outlined in the preceding section. $\square$

Finally, let (M, ω) be an arbitrary Hamiltonian G-space with moment map $\phi : M \to g^*$. Let p be an extremal fixed point of G. Near the point p the action of G is isomorphic to the linear isotropy action of G on the tangent space T_p by the equivariant Darboux theorem. Therefore, taking $V = T_p$, the conclusions of the preceding theorem are valid for regular values, λ, of ϕ providing that λ is sufficiently close to $\phi(p)$. This proves Theorem 9.8.3. In particular, for such values λ we can compute the cohomology ring of X_λ by Theorem 9.8.6.

9.9   Equivariant Duistermaat-Heckman

Let G and K be tori, and let (M, μ) be a Hamiltonian $(G \times K)$-space with moment map

$$(\phi, \psi) : M \to g^* \oplus k^*,$$

and let

$$\tilde{\mu} \in \Omega^2_{G \times K}(M)$$

be the corresponding $G \times K$ equivariant symplectic form .

Let a be a regular value of ϕ and define

$$Z_a := \phi^{-1}(a), \quad X_a := Z_a/G.$$

Let

$$i : Z_a \to M$$

be the inclusion and

$$\pi : Z_a \to X_a$$

the projection. The group K acts on these spaces.

The equivariant version of the Marsden-Weinstein reduction theorem asserts that

Theorem 9.9.1 Equivariant Marsden-Weinstein.   *There exists a K equivariant symplectic form*

$$\tilde{\mu}_a \in \Omega^2_K(X_a)$$

such that

$$\pi^* \tilde{\mu}_a = i^* r(\tilde{\mu}) \tag{9.57}$$

where

$$r : \Omega^*_{G \times K}(M) \to \Omega^*_K(M)$$

is the "forgetful" map, i.e. the map corresponding to the inclusion $K \to G \times K$.

Suppose that ϕ is proper. Let $\xi_1, \ldots, \xi_n$ be a basis of g, and $\tilde{c}_1, \ldots, \tilde{c}_n$ the equivariant Chern classes associated with the fibration

$$
\begin{array}{ccc}
G & \hookrightarrow & Z_a \\
 & & \downarrow \pi \\
 & & X_a
\end{array}
$$

The equivariant version of the Duistermaat-Heckman theorem asserts the following:

Theorem 9.9.2 Equivariant Duistermaat-Heckman. *There exists a neighborhood U of the origin in g^* such that for all $\epsilon \in U$,*

$$X_{a+\epsilon} \approx X_a$$

as K-manifolds, and

$$[\tilde{\mu}_{a+\epsilon}] = [\tilde{\mu}_a] + \sum \epsilon^i \tilde{c}_i, \tag{9.58}$$

where $\epsilon = \sum \epsilon^i \xi_i$.

Here is an important application of this formula. Let m_a be the Duistermaat-Heckman measure on k^* associated with the action of K on X_a. As this measure is compactly supported, it has a well defined Fourier transform

$$\hat{m}(\eta) := \int_{k^*} e^{is(\eta)} dm_a(s) \tag{9.59}$$

which is an analytic function on k. From the definition of m_a, we may write this function as

$$\eta \mapsto \int_{X_a} \exp(\tilde{\mu}_a)(\eta).$$

More generally, we may allow a to vary in a small neighborhood and use (9.58) to write

$$\hat{m}_{a+\epsilon}(\eta) = \int_{X_a} \exp \tilde{\mu}_a(\eta) \exp \left(\sum \epsilon^i \tilde{c}_i(\eta) \right). \tag{9.60}$$

As in section 9.8 we can use this to evaluate the equivariant characteristic numbers

$$[\int_{X_a} p(\tilde{c}_1, \ldots, \tilde{c}_n)(\exp \tilde{\mu}_a)](\eta),$$

where p is a polynomial in n-variables: just apply the differential operator

$$p(\frac{\partial}{\partial \epsilon^1}, \ldots, \frac{\partial}{\partial \epsilon^1})$$

to (9.60) and set $\epsilon = 0$ to obtain

$$p(\frac{\partial}{\partial \epsilon^1}, \ldots, \frac{\partial}{\partial \epsilon^1}) \hat{m}_{a+\epsilon}\Big|_{\epsilon=0} = \int_{X_a} p(\tilde{c}_1, \ldots, \tilde{c}_n)(\exp \tilde{\mu}_a). \tag{9.61}$$

As a special case of this formula we get an interesting equivariant analogue of (9.51). Let $\Delta = \Delta_\lambda$ be the convex polytope (9.44) and let $\hat{v}(\lambda, \eta)$ be the integral of $e^{i\eta s}$ over Δ_λ with respect to the standard Lebesgue measure, ds. Then by theorem (9.8.4), completed with the identity above,

$$\frac{\partial}{\partial \lambda^\beta} \hat{v}(\lambda, \eta) = \int_{X_\Delta} \tilde{c}_1^{\beta_1} \cdots \tilde{c}_d^{\beta_d} \exp \tilde{\mu}.$$

This formula was used in [Gu2] to compute the equivariant Riemann-Roch number of X_Δ and thereby obtain a generalized "Euler-Maclaurin" formula for the sum

$$\sum e^{i\eta K}, \qquad K \in L^* \cap \Delta.$$

(See also [CS].)

9.10   Group Valued Moment Maps

Let G be a compact Lie group and suppose that we put a G invariant scalar product $(\ ,\ )$ on its Lie algebra g. (In case G is simple, this scalar product is unique up to positive multiple.) Let $\theta \in \Omega^1(G, g)$ denote the left invariant Maurer-Cartan form. Then it is well known that the three-form

$$\chi := \frac{1}{12}(\theta, [\theta, \theta])$$

is closed and bi-invariant. (We will review the proof of this fact below.)

It has recently been observed cf. [AMM] that there is a equivariant version of this three-form, i.e. an equivariant three-form $\chi_G \in \Omega_G(G)$ relative to the conjugation action of G on itself which is d_G closed. We shall describe this below.

Suppose that (M, ω, ϕ) is a Hamiltonian G-space. The scalar product gives an isomorphism of $g^* \to g$, and composing this isomorphism with the moment map $\phi : M \to g^*$ we obtain a map $\Phi : M \to g$ which we may also call the moment map. We have the exponential map

$$\exp : g \to G$$

which is G-equivariant for the adjoint action of G on g and the conjugation action of G on itself. This map is a diffeomorphism in a neighborhood of the origin. we can form the composite

$$\nu := \exp \circ \Phi : M \to G.$$

So long as Φ takes values in a neighborhood of the origin where exp is a diffeomorphism, we can translate properties of the moment map Φ into properties of ν and vice versa. (For example, adjoint orbits go into conjugacy classes.) These translations of properties of Φ turn out to involve the equivariant form χ_G mentioned above. But these properties make sense in their own right, and are the subject of study of the recent paper by Alekseev, Malkin and Meinrenken [AMM]where many important applications of these "group valued moment maps" are given. This section consists of an introduction to their paper.

In most of what follows the group G need not be compact and the form (,) need not be positive definite, only non-singular.

9.10.1   The Canonical Equivariant Closed Three-Form on G

Suppose that the Lie algebra g possesses an invariant, non-degenerate symmetric bilinear form (,), so

$$(\xi, \eta) = (\eta, \xi), \quad ([\xi, \eta], \zeta) + (\eta, [\xi, \zeta]) = 0 \quad \forall \xi, \eta, \zeta \in g.$$

This means that the trilinear map

$$q : \xi, \eta, \zeta \mapsto (\xi, [\eta, \zeta])$$

is antisymmetric and invariant, i.e.

$$q \in (\wedge g^*)^g .$$

We have, for $v, \zeta, \xi, \eta \in g$, using the invariance of (,) and Jacobi's identity

$$
\begin{aligned}
([v, \zeta], [\xi, \eta]) &= (v, [\zeta, [\xi, \eta]]) \\
&= (v, [[\zeta, \xi], \eta]) + (v, [\xi, [\zeta, \eta]]) \\
&= -([v, \eta], [\zeta, \xi]) + ([v, \xi], [\zeta, \eta]) \\
&= -([v, \eta], [\zeta, \xi]) - ([v, \xi], [\eta, \zeta]).
\end{aligned}
$$

Thus

$$([v, \xi], [\eta, \zeta]) + ([v, \eta], [\zeta, \xi]) + ([v, \zeta], [\xi, \eta]) = 0.$$

We conclude that

$$\mathcal{A}([v, \zeta], [\xi, \eta]) = 0, \quad \forall v, \zeta, \xi, \eta \in g, \tag{9.62}$$

where $\mathcal{A}$ denotes the alternating sum over all permutations of v, ξ, η, ζ.

Let G be a Lie group with Lie algebra g and suppose that the adjoint representation of G leaves (,) invariant (which is automatic for the connected component of G). So

$$q \in (\wedge g^*)^G .$$

Let θ and $\bar{\theta} \in \Omega^1(G, g)$ denote the left and right Maurer-Cartan forms. If L_g and R_g denote right and left multiplication by $g \in G$, then the values of θ and $\bar{\theta}$ at g are given by

$$\theta_g := dL_{g^{-1}} : TG_g \to TG_e, \quad \bar{\theta}_g := dR_{g^{-1}} : TG_g \to TG_e.$$

In any faithful matrix representation

$$\theta = a^{-1} da, \quad \bar{\theta} = da \cdot a^{-1}.$$

At any $a \in G$ we thus have

$$\bar{\theta}_a = \mathrm{Ad}_a (\theta_a). \tag{9.63}$$

Also (or directly from the definitions) for any fixed $b \in G$,

$$A_b^* \theta = \mathrm{Ad}_b \circ \theta, \quad A_b^* \bar{\theta} = \mathrm{Ad}_b \circ \bar{\theta}. \tag{9.64}$$

Here A_b denotes the conjugation action of b on G, so A_b^* denotes pull-back via this action on forms, and Ad_b denotes the adjoint action of b on g. Note that Ad_b is the derivative of A_b at the point e which is fixed by A_b.

The Maurer-Cartan equations say that

$$d\theta = -\frac{1}{2}[\theta, \theta], \quad d\bar{\theta} = \frac{1}{2}[\bar{\theta}, \bar{\theta}].$$

In particular, the three form

$$\chi := \frac{1}{12}(\theta, [\theta, \theta]) = \frac{1}{12}(\bar{\theta}, [\bar{\theta}, \bar{\theta}]) \tag{9.65}$$

is bi-invariant. It is also closed since

$$\chi = -\frac{1}{6}(\theta, d\theta)$$

by the Maurer-Cartan equations and hence

$$d\chi = -\frac{1}{6}(d\theta, d\theta) = -\frac{1}{24}([\theta, \theta], [\theta, \theta])$$

which vanishes by (9.62). In other words

$$d\chi = 0. \tag{9.66}$$

It is called the canonical three form of G — canonical relative to the choice of $(\,,\,)$. We can extend this to an equivariantly closed three form on G relative to the adjoint action as follows: For any $\xi \in g$, let ξ_G denote the vector field which is the infinitesimal generator of conjugation by $\exp t\xi$. So

$$\xi_G = \xi_R - \xi_L$$

where ξ_R is the right invariant vector field corresponding to ξ (and so is the infinitesimal generator of left multiplication by $\exp t\xi$) and ξ_L is the left invariant vector field (corresponding to right multiplication). Now

$$
\begin{aligned}
\iota(\xi_R)(\bar{\theta}, [\bar{\theta}, \bar{\theta}]) &= (\xi, [\bar{\theta}, \bar{\theta}]) - (\bar{\theta}, [\xi, \bar{\theta}]) + (\bar{\theta}, [\bar{\theta}, \xi]) \\
&= 3(\xi, [\bar{\theta}, \bar{\theta}]) \\
&= 6(\xi, d\bar{\theta}) \\
&= 6d(\bar{\theta}, \xi).
\end{aligned}
$$

Similarly,

$$
\iota(\xi_L)(\theta, [\theta, \theta]) = -6d(\theta, \xi).
$$

Thus

$$
\iota(\xi_G)\chi = \frac{1}{2}d(\theta + \bar{\theta}, \xi). \tag{9.67}
$$

Recall from Section 2.1 that our notational convention in this book is to let ι_ξ denote the interior product with respect to the infinitesimal generator of the one parameter group $\exp(-t\xi)$, in out case the conjugation action of this one parameter group. Hence we can write (9.67) as

$$
\iota_\xi \chi = -\frac{1}{2}d(\theta + \bar{\theta}, \xi). \tag{9.68}
$$

The map of $g \to \Omega(G)$ given by

$$
\xi \mapsto (\theta + \bar{\theta}, \xi)
$$

is G-equivariant. Indeed,

$$
\begin{aligned}
(\theta + \bar{\theta}, \mathrm{Ad}_a \xi) &= \left(\mathrm{Ad}_a^{-1}(\theta + \bar{\theta}), \xi\right) \\
&= \left(A_a^{-1*}(\theta + \bar{\theta}), \xi\right) \quad \text{by (9.64)} \\
&= (A_a^{-1})^*(\theta + \bar{\theta}, \xi).
\end{aligned}
$$

We may therefore define the equivariant three-form

$$
\chi_G \in \Omega_G^3(G), \quad \chi_G(\xi) := \chi - \frac{1}{2}(\theta + \bar{\theta}, \xi). \tag{9.69}
$$

At any $a \in G$ we have

$$
(\iota(\xi_G)\theta)(a) = \mathrm{Ad}_a^{-1}\xi - \xi, \quad (\iota(\xi_G)\bar{\theta})(a) = \xi - \mathrm{Ad}_a\xi
$$

hence

$$
\iota(\xi_G)(\theta + \bar{\theta})(a) = \left(\mathrm{Ad}_a^{-1} - \mathrm{Ad}_a\right)\xi. \tag{9.70}
$$

In particular,

$$
\iota_\xi(\theta + \bar{\theta}, \xi)(a) = -\iota(\xi_G)(\theta + \bar{\theta}, \xi)(a) = -(\mathrm{Ad}_a\xi - \mathrm{Ad}_{a^{-1}}\xi, \xi) = 0
$$

and conclude that

$$
d_G\chi_G = 0. \tag{9.71}
$$

9.10.2   The Exponential Map

For each $s \in \mathbf{R}$ let

$$\exp_s : g \to G$$

be defined by

$$\exp_s(\eta) = \exp(s\eta)$$

where $\exp : g \to G$ denotes the exponential map. Thus $\exp_s(\eta)$ is given as the unique solution of the differential equation with initial conditions

$$\exp_s(\eta)^{-1} \frac{\partial}{\partial s} \exp_s(\eta) = \eta, \quad \exp_0(\eta) = e. \tag{9.72}$$

If A_a denotes the conjugation action of $a \in G$ on G, and Ad_a denotes the adjoint action of a on g, then we have the equivariance condition

$$\exp_s(\mathrm{Ad}_a \eta) = A_a(\exp_s \eta). \tag{9.73}$$

Consider the two form on g defined by

$$\Upsilon := \frac{1}{2} \int_0^1 (\exp_s^* \bar{\theta}, \frac{\partial}{\partial s} \exp_s^* \bar{\theta}) ds. \tag{9.74}$$

For $\xi \in g$ let $v_\xi = \xi_g$ denotes the vector field on g corresponding to the adjoint action of G. We claim that Υ satisfies the following three properties:

$$\mathrm{Ad}_a^* \Upsilon = \Upsilon \quad \forall a \in G \tag{9.75}$$

$$d\Upsilon = -\exp^* \chi \tag{9.76}$$

$$\iota(v_\xi)\Upsilon = -d(\cdot, \xi) + \frac{1}{2} \exp^*(\theta + \bar{\theta}, \xi) \quad \forall \xi \in g. \tag{9.77}$$

To prove the first of these equations, observe that $A_a = L_a R_a^{-1}$ where L_a denotes left multiplication by $a \in G$ and R_a denotes right multiplication. Also, $\bar{\theta}$ is invariant under right multiplication. Hence, by (9.73),

$$\begin{aligned}
\mathrm{Ad}_a^* \exp_s^* \bar{\theta} &= \exp_s^* A_a^* \bar{\theta} \\
&= \exp_s^* L_a^* R_{a^{-1}}^* \bar{\theta} \\
&= \exp_s^* L_a^* \bar{\theta} \\
&= \exp_s^* (\mathrm{Ad}_a \bar{\theta}) \\
&= \mathrm{Ad}_a \exp_s^* \bar{\theta}.
\end{aligned}$$

The invariance of $(\ ,\ )$ then proves (9.75). To prove (9.76) we use the Maurer-Cartan equations for $\bar{\theta}$. We have

$$d\Upsilon = \frac{1}{4} \int_0^1 \left(\exp_s^*[\bar{\theta}, \bar{\theta}], \frac{\partial}{\partial s} \exp_s^* \bar{\theta} \right) ds - \frac{1}{4} \int_0^1 \left(\exp_s^* \bar{\theta}, \frac{\partial}{\partial s} \exp_s^*[\bar{\theta}, \bar{\theta}] \right) ds$$

$$= \frac{1}{4} \int_0^1 \left(\exp_s^*[\overline{\theta}, \overline{\theta}], \frac{\partial}{\partial s} \exp_s^* \overline{\theta} \right) ds - \frac{1}{4} \int_0^1 \frac{\partial}{\partial s} \left(\exp_s^* \theta, \exp_s^*[\overline{\theta}, \overline{\theta}] \right) ds$$

$$+ \frac{1}{4} \int_0^1 \left(\exp_s^*[\overline{\theta}, \overline{\theta}], \frac{\partial}{\partial s} \exp_s^* \overline{\theta} \right) ds$$

$$= \frac{1}{2} \int_0^1 \left(\exp_s^*[\overline{\theta}, \overline{\theta}], \frac{\partial}{\partial s} \exp_s^* \overline{\theta} \right) ds - \frac{1}{4} \exp^*(\overline{\theta}, [\overline{\theta}, \overline{\theta}])$$

$$= \frac{1}{2} \cdot \frac{1}{3} \int_0^1 \frac{\partial}{\partial s} \left([\exp_s^* \overline{\theta}, \exp_s^* \overline{\theta}], \exp_s^* \overline{\theta} \right) ds - \frac{1}{4} \exp^*([\overline{\theta}, \overline{\theta}], \overline{\theta})$$

$$= \frac{1}{2} \cdot \frac{1}{3} \int_0^1 \frac{\partial}{\partial s} \exp_s^*(\overline{\theta}, [\overline{\theta}, \overline{\theta}]) ds - \frac{1}{4} \exp^*([\overline{\theta}, \overline{\theta}], \overline{\theta})$$

$$= -\frac{1}{12} \exp^*([\overline{\theta}, \overline{\theta}], \overline{\theta})$$

$$= - \exp^* \chi.$$

We now prove (9.77). We first recall our notation. For any $\xi \in g$, ξ_R denotes the vector field on G which is the infinitesimal generator of left multiplication by $\exp t\xi$. Thus ξ_R is the right invariant vector field corresponding to ξ and

$$\overline{\theta}_a(\xi_R) = \xi \quad \forall a \in G.$$

Let ξ_L denote the tangent at $t = 0$ to right multiplication by $\exp t\xi$, so ξ_L is the left invariant vector field corresponding to ξ and

$$\overline{\theta}_a(\xi_L) = \mathrm{Ad}_a \xi.$$

If we set $a = \exp t\xi$ in (9.73), and differentiate with respect to t at $t = 0$ we get, for $\eta \in g$,

$$d_\eta \exp_s(v_\xi) = \xi_R(\exp s\eta) - \xi_L(\exp s\eta).$$

If we apply $\overline{\theta}$ to both sides we get

$$\left(\iota(v_\xi) \exp_s^* \overline{\theta} \right)(\eta) = \xi - \mathrm{Ad}_{\exp_s \eta}(\xi). \tag{9.78}$$

Also, we may think of $\xi \in g$ as a constant vector field on g generating the flow

$$\eta \mapsto \eta + t\xi.$$

Define

$$\phi_{s,\eta} : g \to G, \quad \phi_{s,\eta}(\zeta) := \exp_s(\eta + \zeta) \exp_{-s}(\eta)..$$

Thus

$$\exp_s(\eta + t\xi) = \phi_{s,\eta}(t\xi) \exp_s(\eta).$$

Then

$$\iota(\xi) \exp_s^* \overline{\theta} = d_\eta(R_{\exp_s \eta} \circ \exp_{s\eta})(\xi) = \frac{\partial}{\partial t} \phi_{s,\eta}(t\xi)_{|t=0}.$$

On the other hand, the differential equation satisfied by the exponential at $\eta + t\xi$ is

$$\exp_s(\eta + t\xi)^{-1} \frac{\partial}{\partial s} \exp_s(\eta + t\xi) = \eta + t\xi$$

which translates into

$$\exp_s(\eta)^{-1} \phi_{s,\eta}(t\xi)^{-1} \frac{\partial}{\partial s} \left(\phi_{s,\eta}(t\xi) \exp_s(\eta) \right) = \eta + t\xi.$$

Let us differentiate this identity with respect to t and set $t = 0$. The right hand side gives ξ. Applying Leibniz's rule, we get a sum of two terms, the first coming from the t dependence in $\phi(t\xi)^{-1}$ gives

$$- \exp_s(\eta)^{-1} \left(\iota(\xi) \exp_s^* \overline{\theta} \right) \frac{\partial}{\partial s} (\exp_s \eta).$$

The second, coming from the t dependence in the term $\frac{\partial}{\partial s} \left(\phi_{s,\eta}(t\xi) \exp_s(\eta) \right)$ gives

$$\exp_s(\eta)^{-1} \frac{\partial}{\partial s} \left(\iota(\xi) \exp_s^* \overline{\theta} \exp_s(\eta) \right).$$

Expanding the partial derivative with respect to s into two terms by Leibniz's rule again gives two terms, one of which cancels the preceding and we are left with

$$\mathrm{Ad}_{\exp_s(\eta)^{-1}} \frac{\partial}{\partial s} \left(\iota(\xi) \exp_s^* \overline{\theta} \right) = \xi.$$

Applying $\mathrm{Ad}_{\exp_s(\eta)}$ and integrating from 0 to s gives

$$\iota(\xi) \exp_s^* \overline{\theta}(\eta) = \int_0^s \mathrm{Ad}_{\exp_u \eta}(\xi) du. \tag{9.79}$$

We can now prove (9.77): We have

$$
\begin{aligned}
\iota(v_\xi)\Upsilon &= \frac{1}{2} \int_0^1 \left(\iota(v_\xi) \exp_s^* \overline{\theta}, \frac{\partial}{\partial s} \exp_s^* \overline{\theta} \right) ds \\
&\quad - \frac{1}{2} \int_0^1 \left(\exp_s^* \overline{\theta}, \frac{\partial}{\partial s} \iota(v_\xi) \exp_s^* \overline{\theta} \right) ds \\
&= \frac{1}{2} \int_0^1 \left(\iota(v_\xi) \exp_s^* \overline{\theta}, \frac{\partial}{\partial s} \exp_s^* \overline{\theta} \right) ds \\
&\quad - \frac{1}{2} \int_0^1 \frac{\partial}{\partial s} \left(\exp_s^* \overline{\theta}, \iota(v_\xi) \exp_s^* \overline{\theta} \right) ds \\
&\quad + \frac{1}{2} \int_0^1 \left(\iota(v_\xi) \exp_s^* \overline{\theta}, \frac{\partial}{\partial s} \exp_s^* \overline{\theta} \right) ds \\
&= \int_0^1 \left(\iota(v_\xi) \exp_s^* \overline{\theta}, \frac{\partial}{\partial s} \exp_s^* \overline{\theta} \right) ds - \frac{1}{2} \left(\exp^* \overline{\theta}, \iota(v_\xi) \exp^* \overline{\theta} \right).
\end{aligned}
$$

We now apply (9.78) to both terms, to the integrand for all $0 \le s \le 1$ and for the second term with $s = 1$. The second term becomes, when evaluated at $\eta \in g$,

$$\frac{1}{2}(\exp^* \overline{\theta}_\eta, \xi - \mathrm{Ad}_{\exp \eta} \xi) = \frac{1}{2} \exp^*(\overline{\theta} - \theta, \xi)(\eta).$$

The integral becomes

$$\int_0^1 \left(\xi - \mathrm{Ad}_{\exp_s(\cdot)} \xi, \frac{\partial}{\partial s} \exp_s^* \overline{\theta}\right) ds = \int_0^1 \frac{\partial}{\partial s} (\xi, \exp_s^* \overline{\theta}) ds$$

$$- \int_0^1 (\xi, \mathrm{Ad}_{\exp_s(\cdot)^{-1}} \frac{\partial}{\partial s} (\xi, \exp_s^* \overline{\theta})) ds$$

$$= (\exp^* \overline{\theta}, \xi) - d(\cdot, \xi).$$

(The d in the last expression serves to remind us that we are to identify elements of g with constant vector fields.) Adding the two terms gives (9.77). □

9.10.3 G-Valued Moment Maps on Hamiltonian G-Manifolds

Now let M be a Hamiltonian G-manifold with symplectic form ω and with moment map $\phi : M \to g^*$. Composing ϕ with the identification of g^* with g provided by (,), we obtain a map $\Phi : M \to g$ which may then be composed with the exponential map. That is, we may consider the map

$$\nu : M \to G, \qquad \nu := \exp \circ \Phi. \tag{9.80}$$

Clearly ν is an equivariant map for the conjugation action of G on itself. Define the two form Ξ on M by

$$\Xi := \omega + \Phi^* \Upsilon. \tag{9.81}$$

The form Ξ is invariant under G since ω and Υ are. To compute $d\Xi$ we use the fact that $d\omega = 0$ and (9.76). We get

$$d\Xi = -\nu^* \chi. \tag{9.82}$$

Let ξ_M^- denote the vector field on M corresponding to $\xi \in g$, so that the moment map property of Φ says that

$$\iota_\xi \omega = \iota(\xi_M^-)\omega = -d(\Phi, \xi).$$

Using (9.77) we obtain

$$\iota(\xi_M)\Xi = d(\Phi, \xi) + \frac{1}{2}\Phi^* \exp^*(\theta + \overline{\theta}, \xi) - d(\Phi, \xi)$$

or

$$\iota(\xi_M)\Xi = \frac{1}{2}\nu^*(\theta + \overline{\theta}, \xi), \qquad \forall \xi \in g. \tag{9.83}$$

We can combine equations (9.82) and (9.83) as follows: Consider Ξ as an element of $\Omega_G^2(M)$. Then

$$d_G\Xi = -\nu^*\chi_G, \tag{9.84}$$

So far we have not made use of the non-degeneracy of ω. Let us now use this property in computing the kernel of Ξ. Suppose that $v \in TM_x$ satisfies

$$\iota(v)\Xi_x = 0.$$

This means that

$$\iota(v)\omega_x = -\iota(v)\left(\Phi^*\Upsilon\right)_x.$$

The right hand side is an element of T^*M_x which vanishes on all elements of $\ker d\Phi_x$. So

$$v \in (\ker d\Phi_x)^{\omega_x} = (\ker d\phi_x)^{\omega_x},$$

the symplectic orthogonal complement with respect to ω_x of $\ker d\Phi_x$. But one of the basic properties of the moment map (cf. [GS] p.184) is that this orthogonal complement consists precisely of the evaluation of vector fields coming from the G-action. In other words,

$$v = \xi_M(x)$$

for some $\xi \in g$. So, by (9.83), we conclude that

$$\left(\Phi^* \exp^*(\theta + \overline{\theta}, \xi)\right)_x = 0.$$

Now for any $\eta \in g$, $d\Phi(\eta_M) = \eta_g$ and $d\exp\eta_g = \eta_G$. Hence, by (9.70), if we take the interior product of the preceding equation with $\eta_M(x)$ we get

$$\left(\eta, (\mathrm{Ad}_{\nu(x)}^{-1} - \mathrm{Ad}_{\nu(x)})\xi\right) = 0.$$

Since this must hold for all $\eta \in g$, we conclude that

$$\xi \in \ker\left(\mathrm{Ad}_{\nu(x)}^2 - 1\right) = \ker\left(\mathrm{Ad}_{\nu(x)} - 1\right) \oplus \ker\left(\mathrm{Ad}_{\nu(x)} + 1\right). \tag{9.85}$$

Conversely, if ξ lies in this kernel, then $(\iota(\xi_G)(\theta + \overline{\theta}))(\nu(x)) = 0$ by (9.70), and hence $\xi_M(x)$ lies in the kernel of Ξ by (9.83).

Let us show that if $\Phi(x)$ is sufficiently close to the origin, there are no non-zero elements ξ of the first summand on the right of (9.85) such that $\xi_M(x)$ lies in the kernel of Ξ. Indeed, if $\mathrm{Ad}_{\nu(x)}\xi = \xi$, then $\iota(\xi_G)_{\nu(x)}\overline{\theta} = 0$ and our condition becomes

$$\left(\Phi^*(\exp^*\theta, \xi)\right)_x = 0.$$

Suppose that the differential of the exponential map is bijective at $\Phi(x) \in g$. Then this last condition says that ξ annihilates the image of the differential of the moment map. But, by another basic property of the moment map, (cf. [GS] page 184 again) this implies that $\xi_M(x) = 0$.

We have thus proved that

$$\ker \Xi_x = \{\xi_M(x), \;\; \xi \in \ker \left(\mathrm{Ad}_{\nu(x)} + 1\right)\} \tag{9.86}$$

if $\Phi(x)$ is a regular point of the exponential map. This leads Alekseev, Malkin and Meinrenken [AMM] to make the following definition:

Definition 9.10.1 *A G-manifold M together with a G-invariant two form Ξ and a G-equivariant smooth map $\nu : M \to G$ is called a q-Hamiltonian G-space if (9.84) and (9.86) hold, i.e. if*

- $d_G\Xi = -\nu^*\chi_G$ *   and*

- $\ker \Xi_x = \{\xi_M(x)| \;\; \xi \in \ker \left(Ad_{\nu(x)} + 1\right)\}.$

We can summarize the preceding discussion (and read it backwards for the converse) as

Proposition 9.10.1 *Let M be a Hamiltonian G-space with moment map Φ considered as a map $\Phi : M \to g$. Suppose that the image of Φ consists of regular points for the exponential map. Then*

$$\nu = \exp \circ \Phi$$

and

$$\Xi := \omega + \Phi^*\Upsilon$$

give M the structure of a q-Hamiltonian G-space. Conversely, suppose M is a q-Hamiltonian G-space such that $\nu(M)$ lies in $\exp U$ where U is a neighborhood of the origin for which the exponential map is a diffeomorphism. Then $\Phi := \exp^{-1} \circ \nu$ and the preceding equation make M into a Hamiltonian G-space.

9.10.4   Conjugacy Classes

Let C be a conjugacy class of G. We will show that C carries a canonical invariant two-form which makes C into a q-Hamiltonian G-space relative to the canonical embedding $\nu : C \to G$ of C as a submanifold of G. We first make a preliminary remark. If a and b are elements of G, then $A_a b = b$ if and only if $A_b a = a$. Hence, for $\xi \in g$,

$$\xi_G(a) = 0 \leftrightarrow \mathrm{Ad}_a \xi = \xi$$

The tangent space to C at any $a \in C$ consists of the vectors

$$\xi_G(a), \quad \xi \in g.$$

We define

$$\Xi(\xi_G(a), \eta_G(a)) := \frac{1}{2}\left((\eta, \operatorname{Ad}_a \xi) - (\xi, \operatorname{Ad}_a \eta)\right) = \frac{1}{2}\left((\operatorname{Ad}_a^{-1} - \operatorname{Ad}_a)\eta, \xi\right).$$
(9.87)

This is well-defined since if $\eta_G(a) = 0$, then $\operatorname{Ad}_a \eta = \eta = \operatorname{Ad}_a^{-1}\eta$ by the above remark.

It follows from (9.70) that

$$\Xi(\xi_G(a), \eta_G(a)) = \frac{1}{2}\left((Ad_a^{-1} - Ad_a)\eta, \xi\right) = \frac{1}{2}\iota(\eta_G(a))(\theta_a + \overline{\theta}_a, \xi).$$

Thus

$$\iota(\xi_G)\Xi = \frac{1}{2}\nu^*(\theta + \overline{\theta}, \xi).$$

So to verify (9.84) we must show that

$$d\Xi = -\nu^*\chi.$$

For this, consider the map

$$\rho_a : G \to C, \quad \rho_a(b) := A_b a = bab^{-1}.$$

Then

$$\rho_a(u \cdot \exp t\xi) = (u \exp t\xi \cdot u^{-1})uau^{-1}\left(u \exp t\xi \cdot u^{-1}\right)^{-1}$$

so

$$d(\rho_a)_u(\xi_L) = (\operatorname{Ad}_u \xi)_G.$$

Since

$$(\operatorname{Ad}_{uau^{-1}} \operatorname{Ad}_u \xi, \operatorname{Ad}_u \eta) = (\operatorname{Ad}_u \operatorname{Ad}_a \xi, \operatorname{Ad}_u \eta) = (\operatorname{Ad}_a \xi, \eta)$$

we see that

$$\rho_a^*\Xi = \frac{1}{2}(\operatorname{Ad}_a \theta, \theta),$$

hence

$$\rho_a^* d\Xi = d\rho_a^*\Xi = -\frac{1}{4}(\operatorname{Ad}_a[\theta, \theta], \theta) + \frac{1}{4}(\operatorname{Ad}_a \theta, [\theta, \theta]),$$

a left-invariant three-form. It will suffice to show that this equals $-\rho_a^*\nu^*\chi$. For this observe that at any $u \in G$ we have

$$\begin{aligned}
d(\rho_a)_u(\xi_L(u)) &= (\operatorname{Ad}_u \xi)_G(uau^{-1}) \\
&= (\operatorname{Ad}_u \xi)_R(uau^{-1}) - (\operatorname{Ad}_u \xi)_L(uau^{-1}) \\
&= (\operatorname{Ad}_{ua^{-1}} \xi)_L(uau^{-1}) - (\operatorname{Ad}_u \xi)_L(uau^{-1})
\end{aligned}$$

so

$$\rho_a^*\nu^*\theta = \operatorname{Ad}_u\left(\operatorname{Ad}_{a^{-1}}\theta - \theta\right).$$

In computing $\rho_a^* \nu^* \chi$ at u we can drop the Ad_u occurring to the left in this expression by the invariance of (,). Thus

$$\rho_a^* \nu^* \chi = \frac{1}{12} \left((\mathrm{Ad}_{a^{-1}} \theta - \theta), [(\mathrm{Ad}_{a^{-1}} \theta - \theta), (\mathrm{Ad}_{a^{-1}} \theta - \theta)] \right).$$

Expanding out the terms and using the invariance of (,) completes the verification. We have thus established (9.84).

To establish (9.86), suppose that $\xi_G(a) \in \ker \Xi_a$. This means that

$$\mathrm{Ad}_a \, \xi - \mathrm{Ad}_{a^{-1}} \, \xi = 0,$$

or

$$\xi \in \ker \left(\mathrm{Ad}_a^2 - 1 \right) = \ker \left(\mathrm{Ad}_a + 1 \right) \oplus \ker \left(\mathrm{Ad}_a - 1 \right).$$

But for $\xi \in \ker (\mathrm{Ad}_a - 1)$ we have $\xi_G(a) = 0$. Thus (9.86) holds. $\square$

9.11   Bibliographical Notes for Chapter 9

1. An action of G on a symplectic manifold, M, is called Hamiltonian if there exists an equivariant moment map, $\phi : M \longrightarrow g^*$, having the properties described in Section 9.1. A necessary condition for an action of G on M to be Hamiltonian is that the symplectic form, ω, be G invariant; however this is far from sufficient. A number of sufficient conditions for a G action to be Hamiltonian are described in [GS], Section 26. For instance if G is compact (as we have been assuming in this monograph) a G action on M is Hamiltonian if either M is compact, or $H^2(M, \mathbf{R}) = 0$ or G is semi-simple.

2. Berline and Vergne are, as far as we know, the first persons to make the observation that a G-action on M is Hamiltonian if and only if ω is the "form part" of an equivariant symplectic form. This observation plays an essential role in their beautiful proof of the Duistermaat-Heckmann theorem in [BV]. (We will describe this proof in Section 10.9.)

3. The classification of homogeneous symplectic manifolds in terms of coadjoint orbits is due to Kostant [Ko1]; however, the quantum version of this result was, in some sense, first observed by Kirillov. Namely, Kirillov proved that if G is a connected unipotent Lie group there is a one-one correspondence between irreducible unitary representations of G and coadjoint orbits. This result was subsequently extended by Kostant [Ko2], Kostant-Auslander [AK], Sternberg [St1], Duflo [Du] et al. (e.g., [Zi], [Li]) to other classes of Lie groups as well. The arsenal of techniques which are used for associating unitary representations to coadjoint orbits are known collectively as "geometric quantization theory" (see [Wo]). To a large extent these techniques are due to Kirillov [Ki], Kostant [Ko2] and Souriau [So].

4. One important example of minimal coupling is the following: Let Y be a manifold and $\pi : P \longrightarrow Y$ a principal G-bundle. Let $X = T^*Y$, and let M be the fiber product of X and P (as fiber bundles over Y). M is a principal G bundle with base, X; and given a connection on P, one can pull it back to M to get a connection on M. For this connection the minimal coupling form (9.11)–(9.12) is symplectic for *all* ϵ. (See [St2].) Moreover, Weinstein [W] observed that there is a way of defining this minimal coupling form intrinsically without recourse to connections: the product, $T^*P \times F$, is a Hamiltonian G-manifold with respect to the diagonal action of G, and its symplectic reduction is symplectomorphic to W with its minimal coupling form.

5. This example of minimal coupling is used in elementary particle physics to describe the "classical" motion of a subatomic particle in the presence of a Yang-Mills field: Suppose that, when the field is absent, this motion is described by a Hamiltonian, $H : T^*Y \longrightarrow \mathbf{R}$. In the presence of a Yang-Mills field (i.e., of a connection on the bundle, $P \longrightarrow Y$) the motion is described by the Hamiltonian, $\rho^* H$, ρ being the fibering of M over T^*Y. (See [St2] and [SU].)

6. Let O be a coadjoint orbit of the group, G, p_0 a fixed base point in O and G_{p_0} the stabilizer group of p_0. If G_{p_0} is compact, there exists a neighborhood, U of p_0 in g^* such that, for every $p \in U$, the coadjoint orbit through p can be reconstructed by a minimal coupling construction in which the base symplectic manifold is O and the fiber symplectic manifold is a coadjoint orbit of G_{p_0}. For some implications of this fact for the representation theory of compact Lie groups see [GLS].

7. For an insightful discussion of minimal coupling from the topological perspective we recommend the paper [GLSW] of Gotay, Lashof, Sniatycki and Weinstein. (They consider the problem which we discuss in Section 9.5, namely the problem of equipping a "twisted product"

$$F \longrightarrow W \longrightarrow X$$

of two symplectic manifolds, F and X, with a symplectic structure, from a more general point of view than ours: They *don't* assume that F is a Hamiltonian G space and that W is of the form $(M \times F)/G$.

8. The Duistermaat-Heckmann theorem described here is one of several versions of Duistermaat-Heckmann (another one of which we will discuss in §10.9). A version of Duistermaat-Heckmann which is easily deducible from Theorem 9.6.3 is the following:

Theorem 9.11.1 *Let (M, ω) be a compact 2d-dimensional Hamiltonian G-manifold with moment map, $\phi : M \longrightarrow g^*$. Then the measure* *on g^* defined by the formula:*

$$\int_{g^*} f \, d\mu_{DH} \;=\; \int_M \phi^* f \omega^d / d!$$

($f : g^* \rightarrow \mathbf{R}$ being an arbitrary continuous function) is piecewise polynomial.

The measure, μ_{DH}, is called the Duistermaat-Heckmann measure. Since it is compactly supported, its Fourier transform:

$$\int e^{ix\xi} \, d\mu_{DH}(\xi) \tag{9.88}$$

is a C^∞ function; and the version of the Duistermaat-Heckmann theorem which we will describe in Chapter 10 is a formula for computing (9.28) at "generic points", $x \in g$.

9. The notion of a q-Hamiltonian G-manifold is an outgrowth of recent attempts to extend various theorems in equivariant symplectic geometry to the action of loop groups on infinite dimensional manifolds. A basic theorem of Alexeev-Malkin-Meinrenken asserts that there is an equivalence of categories between the category of (infinite dimensional) symplectic manifolds equipped with a Hamiltonian loop group action with proper moment maps, and the category of finite dimensional q-Hamiltonian G-manifolds.

10. A beautiful observation of Alexeev-Meinrenken is that there exists an intrinsic volume form on q-Hamiltonian G-manifolds. If dim $M = 2d$, one might regard ω^d as a candidate for such a volume form. However it is in general not non-vanishing. It can be converted into a non-vanishing form by dividing by $\phi^* \chi_\rho$ where χ_ρ is the character of the representation of G whose dominant weight is one-half the sum of the positive roots.

11. An analogue of the equivariant three form in all dimensions has recently been constructed by Alexeev, Meinrenken and Woodward based on an earlier construction of Jeffrey ([Je]).

12. In their study of q-Hamiltonian G-spaces, Alexeev and Meinrenken have been led to consider an entirely new kind of equivariant cohomology in which $(\Omega(M) \otimes S(g^*))^G$ is replaced by $(\Omega(M) \otimes U(g))^G$ where $U(g)$ is the universal enveloping algebra of g. (Recall that $U(g)$ is a filtered algebra and the Poincaré-Birkhhoff-Witt theorem asserts that its associated graded algebra is $S(g)$ which is $\cong S(g^*)$ in the presence of an invariant scalar product.)

Chapter 10

The Thom Class and Localization

Our goal in this chapter is to construct, in a rather canonical way, the equivariant version of the Thom form, following the construction given by Mathai-Quillen [MQ] in the case of ordinary cohomology. We then give some important applications of this construction.

As motivation, we briefly recall the properties of the classical Thom class, referring to Bott-Tu [BT] for details. Let Z be an oriented d-dimensional manifold, and let X be a compact oriented submanifold of codimension k. Then integration over X defines a linear function on $H^{d-k} = H^{d-k}(Z)$ where $H^\ell(Z)$ denotes the de Rham cohomology groups of Z. On the other hand, Poincaré duality asserts that the pairing

$$(\, , \,) : H^{d-k} \times H_0^k \to \mathbf{C}, \quad (a, b) = \int_Z ab$$

is non-degenerate, where $H_0^\ell = H^\ell(Z)_0$ denote the compactly supported-cohomology groups. In particular, there exists a unique cohomology class $\tau(X) \in H^k(Z)_0$ such that

$$\int_X a = (a, \tau(X)) \quad \forall a \in H^{d-k}(Z).$$

This class is called the **Thom class** associated with X, and any closed form τ_X representing this class is called a **Thom form**. Thus a Thom form is a closed form with the "reproducing property"

$$\int_Z \alpha \wedge \tau_X = \int_X \alpha$$

for all closed $\alpha \in \Omega^{d-k}(Z)$. Clearly any closed form with this property is a Thom form.

Although the Thom class is canonically given by Poincaré duality, until relatively recently, the construction of a Thom form involved choices, and hence was not suitable in situations where a more or less canonical construction is necessary for functorial considerations. Furthermore, as we shall see by example at the end of this chapter, Poincaré duality is not available in equivariant cohomology. Both of these problems were solved by the Mathai-Quillen construction whose algebraic aspects we have described in Chapter 7. We will now translate this algebra into geometry.

10.1   Fiber Integration of Equivariant Forms

Let K be a Lie group acting as proper smooth transformations on the oriented manifolds X and Y (preserving the orientations), and suppose that

$$\pi : Y \to X$$

is a fibration which is K-equivariant. Let $m = \dim Y$ and $n = \dim X$ so that

$$k := m - n \geq 0.$$

Let $\Omega^\ell(Y)_0$ denote the space of compactly-supported ℓ-forms on Y with a similar notation for X. If $\ell \geq k$, there is a map

$$\pi_* : \Omega^\ell(Y)_0 \to \Omega^{\ell-k}(X)_0$$

called **fiber integration** where, for $\mu \in \Omega^\ell(Y)_0$, $\pi_*\mu$, is uniquely characterized by

$$\int_Y \pi^*\beta \wedge \mu = \int_X \beta \wedge \pi_*\mu \quad \forall \beta \in \Omega^{m-\ell}(X)_0. \tag{10.1}$$

It is clear that $\pi_*\mu$ is uniquely determined by this condition, since an $(\ell-k)$-form ν on X with the property that

$$\int_X \beta \wedge \nu = 0 \quad \forall \beta \in \Omega^{m-\ell}(X)_0$$

must vanish.

Once we know that our condition determines $\pi_*\mu$ uniquely, it is sufficient (by partitions of unity) to prove the existence in a coordinate patch where $(x^1, \ldots, x^n, t^1, \ldots, t^k)$ are coordinates on Y, where $(x^1, \ldots, x^n)$ are coordinates on X and π is given by $\pi(x, t) = x$. Then if

$$\mu = f_I(x, t) dx^I \wedge dt^1 \wedge \cdots \wedge dt^k + \cdots$$

where the remaining terms involve fewer than k wedge products of the dt^i; we can easily check that

$$\pi_*\mu := \left(\int f_I(x, t) dt^1 \cdots dt^n \right) dx^I$$

satisfies (10.1).

We list some elementary but important properties of fiber integration:

$$d\pi_* = \pi_* d. \tag{10.2}$$

Proof. For any $\mu \in \Omega^{\ell-1}(Y)_0$ and $\beta \in \Omega^{m-\ell}(X)_0$ we have, by two applications of (10.1) and two applications of Stokes,

$$
\begin{aligned}
\int_X \beta \wedge \pi_* d\mu &= \int_Y \pi^* \beta \wedge d\mu \\
&= (-1)^{m-\ell-1} \int_Y \pi^* d\beta \wedge \mu \\
&= (-1)^{m-\ell-1} \int_X d\beta \wedge \pi_* \mu \\
&= \int_X \beta \wedge d\pi_* \mu,
\end{aligned}
$$

so the result follows from the uniqueness of π_*.

If $\phi : Y \to Y$ and $\psi : X \to X$ are proper orientation-preserving maps which are π-related in the sense that $\pi \circ \phi = \psi \circ \pi$ then it follows from the uniqueness of (10.1) that

$$\pi_* \circ \phi^* = \psi^* \circ \pi_*. \tag{10.3}$$

If ϕ_t and ψ_t are one-parameter groups which are π-related, and if v and w are the corresponding vector fields, then v and w are π-related in the sense that

$$d\pi_y(v(y)) = w(\pi(y)).$$

The infinitesimal version of (10.3) then becomes

$$\pi_* \circ L_v = L_w \circ \pi_*, \tag{10.4}$$

where L denotes Lie derivative.

Let v be a vector field on Y and w a vector field on X such that v and w are π-related. We claim that

$$\pi_* i(v) = i(w)\pi_*. \tag{10.5}$$

Indeed, by (10.1), for any $\beta \in \Omega^{m-\ell+1}(X)_0$

$$\int_X \beta \wedge \pi_*(i(v)\mu) = \int_Y \pi^* \beta \wedge i(v)\mu.$$

But

$$\pi^* \beta \wedge \mu = 0$$

since it is an $(m+1)$-form on an m-dimensional manifold. Hence

$$(i(v)\pi^* \beta) \wedge \mu = (-1)^{m-\ell}\pi^* \beta \wedge i(v)\mu.$$

Hence

$$
\begin{aligned}
\int_X \beta \wedge \pi_*(i(v)\mu) &= \int_Y \pi^*\beta \wedge i(v)\mu \\
&= (-1)^{m-\ell} \int_Y (i(v)\pi^*\beta) \wedge \mu \\
&= (-1)^{m-\ell} \int_Y \pi^* (i(w)\beta) \wedge \mu \\
&= (-1)^{m-\ell} \int_X (i(w)\beta) \wedge \pi_*\mu \\
&= \int_X \beta \wedge i(w)\pi_*\mu
\end{aligned}
$$

since $\beta \wedge \pi_*\mu = 0$, as it is an $(n+1)$-form on an n-dimensional manifold. $\square$

From (10.2), (10.4), and (10.5) we see that

Theorem 10.1.1

$$
\pi_* : \Omega(Y)_0 \to \Omega(X)_0
$$

is a morphism of K^* modules.

In particular, we may consider the Cartan complex for both $\Omega(Y)_0$ and $\Omega(X)_0$ and we conclude that

$$
\pi_* d_K = d_K \pi_*. \tag{10.6}
$$

Explicitly, if we regard an element ρ of the Cartan complex of $\Omega(Y)_0$ as a polynomial map $\rho : k \to \Omega(Y)_0$, then $\pi_*(\rho)$ is the polynomial map from k to $\Omega(X)_0$ given by

$$
(\pi_*\rho)(\xi) = \pi_* (\rho(\xi)).
$$

The fact that π_* is a morphism of K^* modules then implies that

$$
\begin{aligned}
(d_K \pi_* \rho)(\xi) &= d\pi_* \rho(\xi) - \iota_\xi \pi_* \rho(\xi) \\
&= \pi_* d\rho(\xi) - \pi_* \iota_\xi \rho(\xi) \\
&= (\pi_* d_K \rho)(\xi).   \quad \square
\end{aligned}
$$

For our next property of fiber integration, observe that left multiplication of forms makes $\Omega(X)_0$ into a module over $\Omega(X)$. Since $\pi^* : \Omega(X) \to \Omega(Y)$ is a homomorphism (in fact an injection), this makes $\Omega(Y)_0$ into an $\Omega(X)$-module as well. Then π_* is a homomorphism of $\Omega(X)$-modules:

$$
\pi_* (\pi^*\beta \wedge \mu) = \beta \wedge \pi_*\mu \quad \forall \beta \in \Omega^r(X). \tag{10.7}
$$

Indeed, for any $\alpha \in \Omega^{m-r-\ell}(X)$

$$
\int_X \alpha \wedge \pi_*(\pi^*\beta \wedge \mu) = \int_Y (\pi^*\alpha \wedge \pi^*\beta) \wedge \mu
$$

$$= \int_Y \pi^* (\alpha \wedge \beta) \wedge \mu$$

$$= \int_X (\alpha \wedge \beta) \wedge \pi_* \mu$$

$$= \int_X \alpha \wedge (\beta \wedge \pi_* \mu)$$

so (10.7) holds.   □

Finally, suppose that we have an exact square of maps

$$
\begin{array}{ccc}
Y_1 & \xrightarrow{\ g\ } & Y \\
{\scriptstyle \pi_1}\downarrow & & \downarrow{\scriptstyle \pi} \\
X_1 & \xrightarrow[\ h\ ]{} & X
\end{array}
$$

such that π_1 is a fibration, the restriction of g to each fiber $\pi_1^{-1}(x_1)$ is a diffeomorphism, and g and h are proper. Then, in generalization of (10.3), we have

$$(\pi_1)_* g^* = h^* \pi_*. \tag{10.8}$$

Proof. For every $y \in Y$ we can find neighborhoods U of y and W of $\pi(y)$ such that the square

$$
\begin{array}{ccc}
U & \xrightarrow{\ i\ } & \mathbf{R}^n \times \mathbf{R}^k \\
{\scriptstyle \pi}\downarrow & & \downarrow{\scriptstyle \mathrm{pr}_1} \\
W & \xrightarrow[\ j\ ]{} & \mathbf{R}^n
\end{array}
$$

is commutative where i and j are open embeddings and pr_1 is projection onto the first factor. Any form ω supported in U can be written in these local coordinates as

$$\omega = a_I dx^I \wedge dt^K + \cdots, \qquad K = (1, \ldots, k)$$

and

$$\pi_* \omega = j^* \left(\int a_I(x, t) dx^I dt^1 \cdots dt^k \right).$$

An element of $g^{-1}(U)$ can be described as (z, t), $z \in h^{-1}(W)$ and

$$(g^* \omega)_{z,t} = a_I(h(z), t) \left(h^* dx^I \right) dt^K + \cdots$$

and hence at z we have

$$(\pi_1)_* g^* \omega = \left(\int a_I(h(z), t) dt^1 \cdots dt^k \right) h^* dx^I = h^* \pi_* \omega.$$

A partition of unity argument then proves (10.8) for all forms of compact support.

We conclude this section by pointing out the relation between fiber integration and Thom forms. Suppose that X is a compact submanifold of a manifold Z, and U is tubular neighborhood of X which we identify with an open neighborhood of X in the normal bundle NX of X. Let $\tau \in \Omega^k(U)_0$ be a closed form, where k is the codimension of X. Then $\pi_*\tau$ is a zero-form, i.e. a function on X. Suppose that $\pi_*\tau = 1$ (that is integrating over every fiber results in the value one). Then τ is a Thom form. Indeed, since X is a deformation retract of U, for any closed form α on Z, the restriction of α to U is cohomologous to $\pi^*i^*\alpha$. Hence, if α is of degree $d - k = \dim X$,

$$
\begin{aligned}
\int_Z \alpha \wedge \tau &= \int_U \alpha \wedge \tau \\
&= \int_U \pi^*i^*\alpha \wedge \tau \\
&= \int_X i^*\alpha \wedge \pi_*\tau \\
&= \int_X \alpha. \quad \square
\end{aligned}
$$

10.2   The Equivariant Normal Bundle

Let Z be an oriented d-dimensional manifold on which K acts, and suppose that X is a compact $(n = d - k)$-dimensional oriented submanifold which is invariant under the action of K. Let

$$ i : X \to Z $$

denote the inclusion map.

Let $N = NX$ denote the normal bundle of X in Z. By the equivariant tubular neighborhood theorem, there exists a K-invariant tubular neighborhood U of X in Z, and a K-equivariant diffeomorphism

$$ \gamma : N \to U $$

such that

$$ \gamma \circ i_0 = i $$

where $i_0 : X \to N$ is the embedding of X into N as the zero section.

Suppose we put a K-invariant scalar product $(\ ,\ )$ on N, which we can always do if K is compact. Let P be the oriented orthonormal frame bundle of N:

$$ P_x = \{ \mathbf{e} = (e_1, \ldots, e_k) | e_i \in N_x, \quad (e_i, e_j) = \delta_{ij} \}. $$

Let M be the compact Lie group $SO(k)$ and $V := \mathbf{R}^k$ its standard module. Then P is a principal M-bundle over X. Because the scalar product $(\ ,\ )$ on N is K-invariant, the map

$$K \times P \to P, \quad (a, (x, \mathbf{e}) \mapsto (ax, (ae_1, \ldots, ae_k))$$

defines an action of K on P which covers its action on X.

We extend this to an action of K on $P \times V$ by letting K act trivially on V. The map

$$P \times V \to N, \quad ((x, \mathbf{e}), \mathbf{a}) \mapsto (x, a_1 e_1 + \cdots + a_k e_k) \qquad (10.9)$$

makes $P \times V$ into a principal M bundle over N, and the actions of K and M on $P \times V$ commute. Thus the map (10.9) descends to give a K-equivariant diffeomorphism of $(P \times V)/M$ with N.

Again, if K is compact, we may put a K-invariant connection on P. We are thus in the situation of Section 4.6 where we have two commuting actions, and $P \times V$ is equipped with connection forms relative to the $SO(k)$-action. We conclude that the Cartan map gives isomorphisms

$$\kappa_N : \Omega_{M \times K}(P \times V)_0 \to \Omega_K(N)_0.$$

and

$$\kappa_X : \Omega_{M \times K}(P)_0 \to \Omega_K(X)_0.$$

Finally, we can consider the equivariant de Rham complex $\Omega_{M \times K}(V)_0$. Since K acts trivially on V, we have

$$\Omega_{M \times K}(V)_0 = \Omega_M(V)_0 \otimes S(k^*)^K.$$

In particular, we have an embedding

$$\Omega_M(V)_0 = \Omega_{SO(k)}(V)_0 \to \Omega_{M \times K}(V)_0, \quad \beta \mapsto \beta \otimes 1.$$

Here is how we will construct our equivariant Thom form: First we will make a slight modification in the universal Thom-Mathai-Quillen form ν, given by (7.16) and (7.19) so as to make it of compact support instead of merely vanishing rapidly at infinity. Let us call this modified form ν_0. So $\nu_0 \in \Omega_M(V)_0$ and hence

$$\nu_0 \otimes 1 \in \Omega_{M \times K}(V)_0.$$

Let pr_2 denote the projection onto the second factor:

$$\mathrm{pr}_2 : P \times V \to V$$

so

$$\mathrm{pr}_2^* : \Omega_{M \times K}(V) \to \Omega_{M \times K}(P \times V).$$

We define

$$\tau := \kappa_N \left(\mathrm{pr}_2^*(\nu_0 \otimes 1) \right) \in \Omega_K(N). \qquad (10.10)$$

We will show that τ, so defined, has all the desired properties of an equivariant Thom form.

10.3   Modifying ν

The form ν given by (7.16) and (7.19) is of the form

$$\nu = e^{-\frac{1}{2}\sum u_i^2}\left(p_I(x)du^I\right).$$

Here the u_i are linear coordinates on V and the x^i are linear coordinates on $m = o(k)$, and the p_I are polynomials in the x and do not depend on the u. Let B denote the open unit ball in V and consider the $SO(V)$-equivariant diffeomorphism

$$\rho: B \to V, \quad u \mapsto \frac{1}{1-\|u\|^2}u.$$

Then

$$\rho^*\left(e^{-\frac{1}{2}\|u\|^2}\right) = e^{-\frac{\|u\|^2}{2(1-\|u\|^2)^2}}$$

vanishes to infinite order as $\|u\| \to 1$ which is enough to kill any growth coming from denominators in ρ^*du^I. Thus if we extend $\rho^*\nu$ by setting it equal to zero outside the unit ball, we obtain a form ν_0 of compact support which is d_M-closed and whose total integral is one.

10.4   Verifying that τ is a Thom Form

We know that $\mathrm{pr}_2^*(\nu_0 \otimes 1)$ is closed form in $\Omega_{M \times K}(P \times V)_0$ and hence from the Cartan isomorphism that τ given by (10.10) is d_K closed. We wish to show that

$$\pi_*\tau = 1. \tag{10.11}$$

Once we do this, notice that if we then identify N with the tubular neighborhood U of X, and consider τ as a form on Z (by extending by zero) then $[\tau]$, the class of τ is a Thom class. Indeed, since X is a deformation retract of U, id_U and $i \circ \pi$ are homotopic. Hence if α is any closed form, its restriction to U is cohomologous to $\pi^*i^*\alpha$. Hence

$$
\begin{aligned}
\int_Z \alpha \wedge \tau &= \int_U \alpha \wedge \tau \\
&= \int_U \pi^*i^*\alpha \wedge \tau \\
&= \int_X i^*\alpha \wedge \pi_*\tau \\
&= \int_X \alpha
\end{aligned}
$$

by (10.11). So we must prove (10.11).

Let us first prove this in the non-equivariant case, i.e. when $K = \{e\}$. Fix some point $x_0 \in X$, and let $P_0 := \pi^{-1}(x_0)$ be the fiber over x_0 and N_0 the fiber of N over x_0. We have the fibrations

$$P_0 \to x_0, \quad P_0 \times V \to N_0$$

the unique connection on P_0 arising from the fact that it is a homogeneous space for $M = SO(k)$, the Cartan map

$$\kappa_0 : \Omega_G(P_0 \times V) \to \Omega(N_0),$$

and the element

$$\tau_0 = \kappa_0 \left(\mathrm{pr}_2^* \nu_0\right).$$

It follows from the functoriality of our constructions that

$$\tau_0 = j_0^* \tau$$

where $j_0 : N_0 \to N$ is the inclusion of N_0 as a fiber of N. So to prove (10.11) in the case of trivial K, it is enough to prove it for the case of a fibration over a point.

Since our connection forms all lie in P_0, in the bundle

$$P_0 \times V \to N_0$$

all the spaces $\{p\} \times V$ are horizontal and all curvature forms vanish. Let $\xi_1, \ldots, \xi_r$ be a basis of $m = o(k)$ and let $x^1, \ldots, x^r$ be the dual basis. In the Cartan model, an equivariant de Rham form on $P_0 \times V$ is a polynomial in $x^1, \ldots, x^r$ with ordinary de Rham forms on $P_0 \times V$ as coefficients. So it is a sum of terms of the form

$$\beta = \beta_I x^I.$$

By Cartan's procedure, we must replace the x^i by the curvature forms μ^i and then take the horizontal component. But the curvatures all vanish, so we are left with the "constant term" β_0. We must take the horizontal component of β_0 which amounts to restricting to each $\{p\} \times V$ which is identified with N_0 via projection. So τ_0 is just the "constant term" in the expansion of ν_0 as a polynomial in x, under the identification of N_0 with V via

$$V = \{p\} \times V \to P_0 \times V \to N_0.$$

(The choice of p is irrelevant, by invariance.) But the integral of this "constant term" is precisely the integral of ν_0 which equals 1. This proves (10.11) for the case that K is trivial.

Let us now prove (10.11) in the general case. We keep the notation $x^1, \ldots, x^r$ for a basis of the dual space to $m = o(k)$ so that

$$\nu_0 = \nu_{0I} x^I$$

and hence
$$\tau = \kappa_N \left(\mathrm{pr}_2^*(\nu_0 \otimes 1) \right) = \left(\mathrm{pr}_2^*(\nu_{0I}) \right)_{\mathrm{hor}} \tilde{\mu}^I$$
where the $\tilde{\mu}^a$ are now the "equivariant curvature forms", that is the curvature forms for $SO(k)$ but computed using the operator d_K instead of d. But
$$\tilde{\mu}^a = \mu^a - \phi^a$$
where the μ^a are the non-equivariant curvature forms and $\phi^a \in k^* \otimes \Omega^0(P)$. Thus
$$\tau = \tau_0 + \tau_j \otimes p^j$$
where τ_0 is the non-equivariant Mathai-Quillen-Thom form with the $\tau_j \in \Omega^{k-2j}(N)_0$ and the $p^j \in S^j(k^*)$. Upon fiber integration, all the terms $\pi_*(\tau_j \otimes p^j) = \pi_* \tau_j \otimes p^j = 0$ since the fiber degree is less that the fiber dimension. $\square$

10.5   The Thom Class and the Euler Class

Suppose that X has even codimension in Z.

Theorem 10.5.1 *Let $i : X \to Z$ be the inclusion map, let $\tau(X)$ be the Thom class of X and $e(N)$ the Euler class of its normal bundle, N. Then*
$$i^*\tau(X) = e(N). \tag{10.12}$$

Proof. We have represented N as the quotient space $N = (P \times V)/M$ so we have the commutative diagram

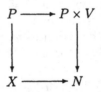

where the horizontal arrows are inclusions and hence the commutative diagram

$$
\begin{array}{ccc}
H^*_{M \times K}(P \times V) & \longrightarrow & H^*_K(N) \\
\downarrow & & \downarrow \\
H^*_{M \times K}(P) & \longrightarrow & H^*_K(X)
\end{array}
$$

where the horizontal arrows are isomorphisms and the vertical arrows are pullbacks. Our construction of the Thom class gives $\tau(X)$ as the image via the top horizontal arrow of
$$\mathrm{pr}_2^*[\nu_0 \otimes 1], \quad \mathrm{pr}_2 : P \times V \to V.$$

So $i^*\tau(X)$ is the image by the bottom horizontal isomorphism of $\mathrm{pr}_2^*[i^*\nu_0 \otimes 1]$ and the result follows from (7.20). $\square$

10.6   The Fiber Integral on Cohomology

Let $N \to X$ be a K-equivariant vector bundle of rank k over a compact manifold X. Since π_* is a cochain map, it induces a map on cohomology

$$\pi_* : H_K^\ell(N)_0 \to H_K^{\ell-k}(X). \tag{10.13}$$

Theorem 10.6.1 *The map (10.13) is an isomorphism.*

Proof. In the case that X is a point, this is the content of (7.21) (except that we used rapidly-vanishing forms instead of compact ones, but this clearly makes no difference). From this it follows that the projection $P \times V \to P$ induces an isomorphism on cohomology

$$H_{SO(k) \times K}^*(P \times V)_0 \to H_{SO(k) \times K}^*(P)$$

where $N = (P \times V)/SO(k)$ as above. Then the square

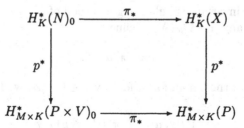

induces a square

$$
\begin{array}{ccc}
H_K^*(N)_0 & \xrightarrow{\ \ \pi_*\ \ } & H_K^*(X) \\
\ \ \downarrow{p^*} & & \ \ \downarrow{p^*} \\
H_{M \times K}^*(P \times V)_0 & \xrightarrow[\ \ \pi_*\ \ ]{} & H_{M \times K}^*(P)
\end{array}
$$

in which the vertical arrows and the bottom horizontal arrow are isomorphisms. Hence the top arrow is an isomorphism.   □

   An important corollary of this result is that the equivariant Thom class $[\tau] \in H_K^k(N)_0$ is uniquely characterized by the property

$$\pi_*[\tau] = 1.$$

10.7   Push-Forward in General

If $\pi : Y \to X$ is a submersion, we have defined π_* on compactly supported forms. One of its key properties is that its induced map on cohomology is a transpose to pullback:

$$\int_X \pi_*[\alpha] \wedge [\beta] = \int_Y [\alpha] \wedge \pi^*[\beta].$$

In ordinary cohomology, where we have Poincaré duality, we could take this as the definition of push-forward, and it would work for any smooth map $f : Y \to X$, not necessarily a submersion. We would like to define push-forward in general for equivariant cohomology as well, and we would like it to have this transpose property. But we can not use this property as a definition, since we do not have Poincaré duality.

Any map $f : Y \to X$ can be factored

$$f = \pi \circ \gamma$$

where

$$\gamma : Y \to Y \times X, \quad \gamma(y) := (y, f(y))$$

is the inclusion into the graph and

$$\pi : Y \times X \to X$$

is projection onto the second factor.

So we need to define the push forward in equivariant cohomology for the case of an equivariant inclusion

$$i : X \to Z,$$

and check that it has the desires adjoint property relative to pullback. Let $\pi : U \to X$ be an invariant tubular neighborhood of X in Z and let $[\tau]$ be its Thom class. For any $\mu \in \Omega_K(X)_0$ define

$$i_* \mu := \pi^* \mu \wedge \tau.$$

By the property of the Thom form, for any $\beta \in \Omega_K(Z)$ we have

$$\int_X i^* \beta \wedge \mu = \int_Z \pi^*(i^* \beta \wedge \mu) \wedge \tau.$$

If β is closed, $\pi^* i^* \beta$ is cohomologous to β and so if μ is also closed,

$$\begin{aligned}
\int_X i^* \beta \wedge \mu &= \int_Z \beta \wedge \pi^* \mu \wedge \tau \\
&= \int_Z \beta \wedge i_* \mu
\end{aligned}$$

which is the desired adjointness property. $\square$

10.8   Localization

Let G be a compact Lie group acting on a compact, oriented, d-dimensional manifold M preserving the orientation. The idea of the localization theorem is to express the integral of an equivariant de Rham form over M in terms of its "residues" at the fixed points of G.

So let X be a connected component of the fixed point set M^G, and suppose we have chosen an orientation on X. Let $\pi : U \to X$ be an invariant tubular neighborhood and $\tau \in \Omega_G(U)_0$ an equivariant Thom form. We know that

$$i^*[\tau] = e(N), \tag{10.14}$$

the Euler class of N, where $N = N_X$ is the normal bundle of X in M. We have observed that the restriction of any closed form $\mu \in \Omega_G(M)$ to U is homologous to $\pi^* i^* \mu$ and hence

$$\int_U \mu \wedge \tau = \int_U \pi^* i^* \mu \wedge \tau = \int_X i^* \mu \wedge \pi_* \tau = \int_X i^* \mu$$

since

$$\pi_* \tau = 1.$$

Suppose that μ is compactly supported in U. We can interchange the roles of τ and μ in the above argument to get

$$\begin{aligned}
\int_X i^* \mu &= \int_U \mu \wedge \tau \\
&= \int_U \mu \wedge \pi^* i^* \tau \\
&= \int_U \mu \wedge \pi^* e(N)
\end{aligned}$$

by (10.14). If the codimension of X is odd, $e(N) = 0$ and the formula says that $\int_X i^* \mu = 0$. The formula becomes more interesting in the case of even codimension when we can write it as

$$\int_X i^* \mu = \int_M \pi^* e(N) \wedge \mu. \tag{10.15}$$

Now G acts trivially on X, since X consists entirely of fixed points. Hence

$$H_G(X) = S(g^*)^G \otimes H(X)$$

so

$$e(N) = f_m + f_{m-1}\alpha_1 + \cdots + \alpha_m$$

where $k = 2m$ and

$$f_i \in S^i(g^*)^G, \quad \alpha_i \in \Omega^{2i}(X)$$

with $d\alpha_i = 0$. Suppose that $f_m \neq 0$ and write the expression for $e(N)$ as

$$e(N) = f_m \left(1 - \frac{\alpha}{f_m}\right), \quad \alpha := -(f_{m-1}\alpha_1 + \cdots + \alpha_m).$$

Formally, let us take the inverse of both sides so as to get

$$\frac{1}{e(N)} = \frac{1}{f_m}\left(1 + \frac{\alpha}{f_m} + \frac{\alpha^2}{f_m^2} + \cdots + \frac{\alpha^{q-1}}{f_m^{q-1}}\right) \tag{10.16}$$

where $q - 1$ is the largest integer $\leq \frac{1}{2}\dim X$. Both sides of (10.16) do not make sense as they stand, but if we multiply the right hand side by f_m^q we get a well defined expression

$$\beta(N) := f_m^{q-1} + f_m^{q-2}\alpha + \cdots + \alpha^{q-1}. \tag{10.17}$$

Since α is d_G-closed, we see that $\beta(N) \in \Omega_G(X)$ is d_G closed, and we have

$$e(N) \times \beta(N) = f_m^q. \tag{10.18}$$

If we replace μ by $\pi^*\beta(N) \wedge \mu$ in (10.15) we get

$$\int_X \beta(N)i^*\mu = f_m^q \int_U \mu. \tag{10.19}$$

This formula expresses an integral over U in terms of an integral over X. We can write it more suggestively as

$$\int_U \mu = \int_X \frac{i^*\mu}{e(N)}. \tag{10.20}$$

The expression $1/e(N)$ on the right hand side of this equation is interpreted as the sum (10.16). So the summands on the right of (10.20) are not elements of $S(g^*)^G$ but expressions of the form

$$\frac{q}{f_m^r}, \quad q \in S(g^*)^G.$$

In other words, they are elements of the ring $S(g^*)^G$ localized at f_m. On the other hand, the left hand side of (10.20) *is* an element of $S(g^*)^G$. So there are a lot of combinatorial cancellations implied by (10.20).

The above localization formula (10.20) was proved under the assumption that $f_m \neq 0$. We now derive some equivalent conditions for this to hold: Let $x_0 \in X$ and let N_0 be the fiber of the normal bundle over x_0. The isotropy representation of G on N_0 preserves the orientation of N_0 (inherited from that of M and X). Since G is compact, we may put a positive definite scalar product on N_0 which is preserved by the isotropy action of G. So we get a group homomorphism

$$G \to SO(N_0) = SO(2m).$$

So we get a transpose map, h, from the ring of invariants $S\left(o(2m)^*\right)^{SO(2m)}$ to $S(g^*)^G$. If j_0 denotes the inclusion of N_0 in N then

$$f_m = j_0^* e(N) = e(N_0).$$

For the case of a point, we have identified the equivariant Euler class as $(2\pi)^{-m}$ times the image under h of the Pfaffian in $S\left(o(2m)^*\right)^{SO(2m)}$. So

$$f_m = (2\pi)^{-m}h(\text{Pfaff}). \qquad (10.21)$$

Let T be a maximal torus of G with Lie algebra t. Any invariant is determined by its restriction to t^*. So we would like to compute the image of $h(\text{Pfaff})$ under the restriction map

$$\lambda : S(g^*) \to S(t^*).$$

For this we use the weight decomposition of N_0 under the isotropy action of the torus: We may identify N_0 with $\mathbf{C}^m$ in such a way that the isotropy representation of T is given by

$$(\exp\xi)(z) = (e^{i\ell_1(\xi)}z_1, \ldots, e^{i\ell_m(\xi)}z_m)$$

where $\xi \in t$ and the ℓ_j in the weight lattice of T. As this representation preserves the standard Hermitian inner product on $\mathbf{C}^m$, we may identify

$$\lambda \circ h(\text{Pfaff})$$

with the restriction to t for the equivariant m-th Chern polynomial of the bundle $N_0 \to x_0$ which is just the product of the roots

$$\ell_1 \cdots \ell_m.$$

We have proved

Theorem 10.8.1 *The restriction of f_m to t is the polynomial $(2\pi)^{-m}\ell_1 \cdots \ell_m$. In particular, $f_m \neq 0$ if and only if the weights of the isotropy representation of T on N_0 are all non-zero.*

Notice that since X is connected, the isotropy representation of T on the fiber of the normal bundle is the same at all points. Also the conjugacy theorem for Cartan subgroups implies that the choice of T is irrelevant. Thus the hypothesis of the theorem is independent of the choice of x_0 and of T. The fact that none of the ℓ_i vanish means that $U^T = X$. So our condition amounts to the assumption that X is a connected component of M^T.

10.9   The Localization for Torus Actions

In this section we will obtain a more global version of (10.20). In view of the fact that (10.20) only holds when $M^G = M^T$ we can, without loss of generality assume that $G = T$, i.e., that G is an r-dimensional torus. To globalize (10.20) we recall a few elementary facts about torus actions on compact manifolds (see Lemma 8.5.1):

Lemma: *Let V be a vector space over $\mathbf{R}$ and $\rho : S^1 \to GL(V)$ a representation of S^1 on V which leaves no vector fixed except 0. Let*

$$A = \frac{d}{dt}\rho(e^{it})|_{t=0}.$$

Then there exists a unique decomposition:

$$V = V_1 \oplus \cdots \oplus V_k$$

and positive integers, $0 < m_1 < \cdots < m_k$ such that $A = m_i J_i$ on V_i with $J_i^2 = -I$.

Corollary: *V admits a canonical complex structure. In particular V is even dimensional and has a canonical orientation.*

Now let G be an r-dimensional torus. Letting $\exp : g \to G$ be the exponential map, there is an exact sequence

$$0 \to \mathbf{Z}_G \to g \xrightarrow{\exp} G \to 0$$

of groups, $\mathbf{Z}_G$ being the *group lattice* of G in g, and its dual, Z_G^*, being the *weight lattice* of G in g^*. We will denote by g_Q the set of elements in g of the form $\eta = \xi/m$ with $\xi \in \mathbf{Z}_G$ and $m \in \mathbf{Z} - 0$. These are precisely the elements in g with the property that *the subgroup of G generated by η is a circle.*

Now let M be a compact, oriented G-manifold. We will need below the following standard result.

Theorem 10.9.1 (Finiteness theorem) *Only a finite number of subgroups, G_i, $i = 1, \ldots, \ell$, can occur as stabilizer groups of points of M.*

Proof. For $p \in M$, let G_p be its stabilizer group, let X be the orbit of G through p, and let N_p be the normal space to X at p. By the equivariant tubular neighborhood theorem

$$(M, X) \cong (NX, X) \tag{10.22}$$

and

$$NX \cong (G \times N_p)/G_p \tag{10.23}$$

and from (10.22) and (10.23) one reads off

Lemma 10.9.1 *There exists a neighborhood U of X in M with the property that the groups which occur as stabilizer groups of points of U are contained in G_p and are identical with the groups that occur as stabilizer groups for the linear action of G_p on N_p.*

This reduces the proof of the finiteness theorem to the proof of the analogous theorem for *linear* actions, (and for linear actions the proof is easy). □

From this finiteness theorem we get as a corollary:

Theorem 10.9.2 *There exist a finite number of weights $\beta_1, \cdots, \beta_q$ in Z_G^* with the property that if $\xi \in g$ satisfies*

$$\beta_i(\xi) \neq 0,\, i = 1, \cdots q \tag{10.24}$$

the corresponding vector field, ξ_M on M is non-zero except at points of M^G.

Proof. If $\xi^{\#}(p) = 0$, ξ has to be in the Lie algebra of the stabilizer group, G_p; therefore for ξ to have the property above it suffices that

$$\xi \notin g_i$$

except when $g_i = g$. However, each of the Lie algebras g_i can be defined by a set of equations of the form, $\beta_1(\xi) = \cdots \beta_k(\xi) = 0$ with β_i's in $\mathbf{Z}_G^*$. $\quad\square$

Corollary 10.9.1 *Let ξ be in g_Q and satisfy (10.24). Then the circle group, S^1, generated by ξ has the property*

$$M^G = M^{S^1}.$$

Let's describe one important consequence of this fact: Consider the set

$$\{\xi \in g,\, \beta_i(\xi) \neq 0,\, i = 1, \ldots, q\}. \tag{10.25}$$

The connected components of this set are called the **action chambers** of g (*vis à vis* the action of G on M).

Proposition 10.9.1 *The connected components of M^G are of even codimension. Moreover, having fixed an action chamber, one can assign to each of these components a canonical orientation.*

Proof. Let $\xi \in g_Q$ be an element of the given action chamber, and S^1 the subgroup of G generated by ξ. Then the connected components of M^G are also connected components of M^{S^1}, and if p is a point on one of these components and N_p the normal space at p, the isotropy representation of S^1 on N_p defines a canonical complex structure, and hence a canonical orientation, on N_p. It's clear moreover, that this orientation doesn't depend on the choice of p but only on the choice of the action chamber in which ξ is contained. $\quad\square$

We will now state and prove the global version of (10.20) which we alluded to above. Let $d = \dim M$ and, for the moment, let $G = S^1$.

Theorem 10.9.3 (Localization theorem for circle actions)
If $\mu \in \Omega_{S^1}^k(M)$, $k \geq d$, is d_{S^1}-closed

$$\int_M \mu = \sum \int_X \frac{i_X^* \mu}{e_X} \tag{10.26}$$

the sum being over the connected components of M^{S^1}, i_X being the inclusion map of X into M and e_X being the equivariant Euler class of the normal bundle of X.

Proof.   Let $U = M - M^{S^1}$. Since S^1 is one-dimensional it acts in a locally free fashion on U; so $H_{S^1}^k(U) = H^k(U/S^1)$ and, in particular, since $\dim U/S^1 = d - 1$, $H_{S^1}^k(U) = 0$ for $k \geq d$. Thus $\mu = d_{S^1}\nu$ on U, for some $\nu \in \Omega_{S^1}^{k-1}(U)$. Let U_X be a tubular neighborhood of X and $\rho_X \in C^\infty(U_X)_0$ a S^1-invariant function which is identically one in a neighborhood of X. Letting $\nu' = \nu - \sum \rho_X \nu$,

$$\mu = d_{S^1}\nu' + \Sigma \mu_X$$

where $\mu_X \in \Omega_G^k(U_X)_0$ and $\mu_X = \mu$ on a neighborhood of X. Thus

$$\int_M \mu = \sum \int_M \mu_X$$

and by (10.20)

$$\int_M \mu_X = \int_X \frac{i_X^* \mu}{e_X}. \qquad \square$$

We'll now extend this result to torus actions. Recall that if K is a subtorus of G there is a natural map

$$H_G(M) \longrightarrow H_K(M) \tag{10.27}$$

induced by the restriction map on forms. More explicitly, given an equivariant form in $\Omega_G(M)$,

$$\mu : g \longrightarrow \Omega(M),$$

one gets an equivariant form in $\Omega_K(M)$ by restricting μ to k, and the restriction, $\mu \longrightarrow \mu_{|k}$, induces the map (10.27) on cohomology. In particular, let $\xi \in g_Q$ and let K be the circle subgroup of G generated by ξ. If ξ satisfies the conditions (10.24) $M^G = M^K$; hence, applying (10.27) to $\mu_{|k}$ and $(e_X)_{|k}$:

$$\int_M \mu(\xi) = \sum \int_X \frac{i_X^* \mu(\xi)}{e_X(\xi)}. \tag{10.28}$$

This identity holds for *all* $\xi \in g_Q$ satisfying the conditions, $\beta_i(\xi) \neq 0, i = 1, \ldots q$. However, g_Q is dense in g. Moreover, the left hand side of (10.28) is a polynomial function of ξ and the terms on the right are rational functions of ξ, so this identity holds, in fact, for all $\xi \in g$ satisfying $\beta_i(\xi) \neq 0$, $i = 1, \ldots q$, or finally, holds for *all* $\xi \in g$, providing one thinks of (10.28) as a formal identity in which the left hand side is an element of the ring $S(g^*)$ and the summands on the right, elements of the ring

$$S(g^*)_f = \left\{ \frac{a}{f^S}, a \in S(g^*), S \geq 0 \right\}$$

for an appropriate $f \in S(g^*) - 0$.

If M^G is finite this localization theorem becomes particularly simple. For $p \in M^G$ let $\alpha_{1,p}, \ldots, \alpha_{m,p}$ be the weights of the isotropy representation of G on T_p. We showed in Section 10.8 that

$$e_p(\xi) = (2\pi)^{-m} \prod_{i=1}^{m} \alpha_{i,p}(\xi)$$

so (10.28) reduces to

$$\int_M \mu(\xi) = (2\pi)^m \sum \frac{i_p^* \mu(\xi)}{\prod \alpha_{i,p}(\xi)} \tag{10.29}$$

for all $\xi \in g$ satisfying

$$\alpha_{i,p}(\xi) \neq 0 \tag{10.30}$$

for all $p \in M$ and $i = 1, \ldots, m$.

Example. Let $\tilde{\omega} = \omega + \phi$ be an equivariant symplectic form with form part $\omega \in \Omega^2(M)^G$ and moment map part $\phi : M \to g^*$. Applying (10.29) to the form

$$\mu = \exp \tilde{\omega} = 1 + \tilde{\omega} + \frac{\tilde{\omega}^2}{2!} + \cdots \tag{10.31}$$

one gets, for every ξ satisfying the conditions (10.30),

$$\int_M e^{\langle \phi, \xi \rangle} \frac{\omega^m}{m!} = (2\pi)^m \sum \frac{e^{\langle \phi(p), \xi \rangle}}{\prod \alpha_{i,p}(\xi)} . \tag{10.32}$$

Remark. Strictly speaking, one can't simply prove this identity by applying (10.29) to the form (10.31) since $\exp \tilde{\omega}$ is not an equivariant de Rham form according to our definition. (It is an analytic function of ξ, but not a *polynomial* in ξ.) However, it is easy to deduce (10.32) from (10.29) by applying (10.29) to each of the terms on the right hand side of (10.32) and summing.

The formula (10.32) is the *Duistermaat-Heckmann theorem.* We will briefly describe how it's related to the Duistermaat-Heckmann theorem of Section 9.7: Let Δ be the image of ϕ. By a theorem of Atiyah and Guillemin-Sternberg, Δ is a convex polytope. Moreover, denoting by g_{reg}^* the set of regular values of ϕ, each connected component of $\Delta \cap g_{reg}^*$ is an open convex subpolytope of Δ. Duistermaat-Heckmann define a measure on the Borel subsets, B, of g^* by setting

$$\nu(B) = \int_{B'} \frac{\omega^m}{m!} , \quad B' = \phi^{-1}(B) .$$

This measure is supported on Δ, and it is easy to show, using elementary measure theory, that it is the product of a smooth function times Lebesgue measure on each of the components of $\Delta \cap g_{reg}^*$. Duistermaat and Heckmann prove, however, a much stronger result: they show that on each connected component of $\Delta \cap g_{\text{reg}}^*$, ν is a *polynomial* function times Lebesgue measure, this polynomial function being the Duistermaat-Heckmann polynomial which we described in Section 9.7. Now the left-hand side of (10.32) can be rewritten in the form,

$$\int e^{\langle \xi, x \rangle} \, d\nu(x)$$

and, replacing ξ by $i\xi$, (10.32) becomes

$$\int e^{i\langle x, \xi \rangle} \, d\nu = (-2\pi i)^m \sum \frac{e^{i\langle \phi(p), \xi \rangle}}{\prod \alpha_{j,p}(\xi)}. \tag{10.33}$$

This formula says that the *Fourier transform of the measure*, ν, is equal to the function on the right hand side at all points, $\xi \in g$, satisfying $\alpha_{j,p}(\xi) \neq 0$ for all j. By inverting this Fourier transform one can get a fairly explicit formula for ν and hence for the Duistermaat-Heckmann polynomials, in terms of the weights $\alpha_{i,p}$. Recall, however, that the Duistermaat-Heckmann polynomials describe how the symplectic volume of the reduced symplectic manifold

$$M_a = \phi^{-1}(a)/G$$

varies as one varies a in a fixed connected component of $\Delta \cap g_{\text{reg}}^*$. Thus (10.32) enables one to express this variation of symplectic volume in terms of the *linear* actions of G in the tangent spaces to M at the fixed points.

10.10   Bibliographical Notes for Chapter 10

1. For a compact oriented G-manifold, M, one has a natural bilinear map of $H_G(M)$ into $S(g^*)^G$ given by the pairing

$$(\mu, \nu) \longrightarrow \int \mu \wedge \nu. \tag{10.34}$$

   This pairing, however, can be highly singular. For instance, if the action of G on M is free the integral over M of a d_G-closed form is zero, so the pairing is trivial. Therefore, one can't define push-forward operations in equivariant de Rham theory simply by invoking Poincaré duality as in the non-invariant case.

2. Some vestiges of Poincaré duality do survive, however, in the equivariant setting. For instance, Ginzburg [Gi] proves that if M is a Hamiltonian G-manifold the pairing (10.34) *is* non-singular on cohomology. (Thus push-forward operations in equivariant cohomology can be defined by Poincaré duality for maps between some G-manifolds.)

3. Our construction of the equivariant Thom form in Section 10.2 is, as we already noted in Chapter 7, due to Mathai and Quillen.

4. In an (unpublished) note, David Metzler describes an alternative way of defining the Thom class and push-forward operations in equivariant cohomology: Given a G-manifold, M, let $\mathcal{E}(M)$ be the complex of de Rham currents. This complex is a G^* module; and if M is oriented, there is a natural inclusion

$$j : \Omega(M) \to \mathcal{E}(M).$$

It is easy to see that this is a morphism of G^* modules, hence it induces a map

$$H_G(\Omega(M)) \to H_G(\mathcal{E}(M)) \qquad (10.35)$$

We claim:

Theorem 10.10.1 *(Metzler) The map (10.35) is bijective.*

Proof. By a well-known theorem of de Rham the map induced by j on ordinary cohomology

$$H(\Omega(M)) \to H(\mathcal{E}(M)) \qquad (10.36)$$

is bijective and hence by Theorem 6.7.1 the map (10.35) is bijective. □

de Rham observed that the isomorphism (10.36) gives one a natural way of defining push-forward operations in ordinary cohomology without explicitly invoking Poincaré duality. The reason for this is that $\mathcal{E}$ is naturally a *covariant* functor with respect to mappings, i.e., if M and N are compact oriented manifolds, a map $f : M \to N$ induces a canonical push-forward on currents

$$f_* : \mathcal{E}(M) \to \mathcal{E}(N) \qquad (10.37)$$

and hence by (10.36) a map

$$f_\# : H(M) \to H(N).$$

If M and N are G-manifolds and f is G-equivariant, it is easy to see that (10.37) is a G^* morphism and, therefore, by the same argument as above, it induces a map on *equivariant* cohomology

$$f_\# : H_G(\mathcal{E}(M)) \to H_G(\mathcal{E}(N))$$

which by (10.35) can be viewed as a map of $H_G(M)$ into $H_G(N)$. In particular, if M is a submanifold of N, and f is the inclusion map, the image under this map of the element $1 \in H_G^0(M)$ is the equivariant Thom class of M.

5. The push-forward operations in equivariant cohomology and the equivariant Thom class can also be defined by purely topological methods, not using de Rham theoretic techniques. (This is, for instance, the approach taken by Atiyah and Bott in §2 of [AB].)

6. The localization theorem was proved independently by Berline-Vergne [BV] and Atiyah-Bott [AB]. The proof that we gave of the localization theorem in this chapter is essentially that of Atiyah and Bott. Let us briefly describe the Berline-Vergne proof, confining ourselves to the case, $G = S^1$ and M^G finite: Let ξ be the infinitesimal generator of the G action and let U be the complement of M^G in M. It is easy to show that there exists a G-invariant one-form, θ, on U with the property, $\iota_\xi \theta = 1$. Consider the equivariant form

$$\nu = \frac{-\theta}{x - d\theta} = \frac{\theta}{x}\left(1 + \frac{d\theta}{x} + \left(\frac{d\theta}{x}\right)^2 + \cdots\right) \tag{10.38}$$

x being the generator of $S(g^*)$. Then

$$d_G\nu = (d_G\theta)(x - d\theta)^{-1} = (x - d\theta)(x - d\theta)^{-1} = 1.$$

From this one concludes that if $\alpha \in \Omega_G(M)$ is d_G-closed

$$d_G(\nu \wedge \alpha) = (d_G\nu) \wedge \alpha = \alpha. \tag{10.39}$$

Let p_r, $r = 1, \ldots, N$, be the fixed points of G and let $z_1, \ldots, z_n$ be a complex coordinate system centered at p_r in which the G-action is the linear action

$$e^{itz} = \left(e^{i\lambda_{1,r}t}z_1, \ldots, e^{i\lambda_{n,r}t}z_n\right)$$

and the vector field, ξ, is

$$\sqrt{-1}\sum \lambda_{k,r}\left(z_k\frac{\partial}{\partial z_k} - \bar{z}_k\frac{\partial}{\partial \bar{z}_k}\right)$$

One can assume without loss of generality that on the set

$$B_r^\epsilon = \left\{z, \Sigma\lambda_{k,r}^2\,|z|^2 < \epsilon^2\right\}$$

the form θ is equal to

$$\frac{1}{2\sqrt{-1}}\frac{\Sigma\lambda_{k,r}(\bar{z}_k\,dz_k - z_k\,d\bar{z}_k)}{\Sigma\lambda_{k,r}^2\,|z_k|^2} \tag{10.40}$$

Let $U^\epsilon = U - \cup B_r^\epsilon$. By (10.39)

$$\int_{U^\epsilon}\alpha = \int_{U^\epsilon}d_G(\nu \wedge \alpha)$$

and hence by Stokes' theorem

$$\int_{U^{\epsilon}} \alpha = \sum_r \int_{\partial B_r^{\epsilon}} \nu \wedge \alpha. \tag{10.41}$$

On B_r^{ϵ}, α is cohomologous to the restriction of α to p_r (viewed as an element, $f_r(x)$, of $S(g^*)$) so there exists an equivariant form, $\beta \in \Omega_G(B_r^{\epsilon})$ satisfying

$$\alpha = f_r(x) + d_G \beta.$$

Hence

$$
\begin{aligned}
\nu \wedge \alpha &= f_r(x)\nu + \nu \wedge d_G \beta \\
&= f_r(x)\nu - d_G(\nu \wedge \beta) + \beta.
\end{aligned}
$$

If we integrate the right hand side over ∂B_r^{ϵ} the second term is zero, and by Stokes' theorem the third term is

$$\int_{B_r^{\epsilon}} d_G \beta$$

which is of order $O(\epsilon)$. Thus

$$\int_{\partial B_r^{\epsilon}} \nu \wedge \alpha = f_r(x) \int_{\partial B_r^{\epsilon}} \nu + O(\epsilon). \tag{10.42}$$

Finally note that by (10.38)

$$\int_{\partial B_r^{\epsilon}} \nu = \frac{1}{x^n} \int_{\partial B_r^{\epsilon}} \theta \wedge d\theta \wedge \cdots \wedge d\theta.$$

Making the change of coordinates,

$$\gamma : \mathbf{C}^n \to \mathbf{C}^n, \gamma(z) = \epsilon(\gamma_{1,r} z_1, \ldots, \gamma_{n,r} z_n).$$

this integral becomes

$$\left(\prod \gamma_{k,r} \right)^{-1} \int_{S^{2n-1}} \theta_o \wedge d\theta_0 \wedge \ldots \wedge d\theta_0 \tag{10.43}$$

where

$$\theta_0 = \frac{1}{\sqrt{-1}} \sum z_k \, d\bar{z}_k - \bar{z}_k dz_k$$

and S^{2n-1} is the standard $(2n-1)$-sphere. By Stokes' theorem

$$\int_{S^{2n-1}} \theta_0 \wedge d\theta_0 \wedge \ldots \wedge d\theta_0 = \int_{B^{2n}} (d\theta_0)^n$$

and the integral on the right is easily computed to be $(2\pi)^n$; so for (10.43) we get

$$(2\pi)^n (\Pi\gamma_{k,r})^{-1}$$

and plugging this into (10.42) and letting ϵ tend to zero we finally obtain the localization formula

$$\int_M \alpha = (2\pi)^n \sum_r \frac{f_r(x)}{x^n} (\Pi\gamma_{k,r})^{-1}$$

which is the same as the formula we derived in §10.9.

7. The localization theorem is a relatively recent result; however, an important special case of this theorem was discovered by Bott in the mid-1960s. Bott's result is concerned with the computation of the characteristic *numbers* of a vector bundle $E \longrightarrow M$ (i.e., the integrals over M of the characteristic classes of E). Suppose that the circle group, $G = S^1$, acts on M and that his action lifts to an action of G on E by vector bundle automorphisms. Then as we pointed out in Chapter 8, there is an equivariant Chern-Weil map

$$S(k^*)^K \longrightarrow H_G(M) \tag{10.44}$$

(with $K = U(n), O(n), \ldots$), and from this one can recover the usual Chern-Weil map by composing it with the restriction map

$$H_G(M) \longrightarrow H(M).$$

Applying the localization theorem to elements in the image of (10.44) one gets polynomial identities in the ring, $S(g^*) = \mathbf{C}[x]$, and by setting $x = 0$ these become "localization" formulas for the usual characteristic numbers. These are the formulas discovered by Bott in the mid-sixties. (See [Bott])

Chapter 11

The Abstract Localization Theorem

In this chapter we will examine the localization theorem from a more abstract perspective and explain why such a theorem "has to be true". As in Section 10.9 we will assume that the group G is a compact connected Abelian Lie group; i.e., an n dimensional torus. The main result of this chapter is a theorem of Borel and Hsiang which asserts that, for a compact G-manifold, M, the restriction map, $H_G(M) \to H_G(M^G)$ is injective "modulo torsion". From this we will deduce a theorem of Chang and Skjelbred which describes the image of this map when M is "equivariantly formal". For this we will need the equivariant versions of some standard results about de Rham cohomology and some elementary commutative algebra. We will go over these prerequisites in Sections 11.1–11.3.

11.1 Relative Equivariant de Rham Theory

Let M be a compact manifold and X a closed submanifold. Denote by $\Omega(M, X)$ the space of de Rham forms whose restriction to X vanishes and by $\Omega(M - X)_c$ the space of compactly supported de Rham forms on $M - X$.

Theorem 11.1.1 *The inclusion map $\Omega(M - X)_c \to \Omega(M, X)$ induces an isomorphism on cohomology.*

This we will deduce from the following:

Lemma 11.1.1 *Let $\pi : U \to X$ be a tubular neighborhood of X in M and let $i : X \to U$ be inclusion. Then if $\omega \in \Omega^k(U)$ is closed and has the property $i^*\omega = 0$, there exists a $\nu \in \Omega^{k-1}(U)$ with the same property and additionally such that $\omega = d\nu$.*

Proof of lemma. X is a deformation retract of U, so the maps $\pi^* : \Omega(X) \to \Omega(U)$ and $i^* : \Omega(U) \to \Omega(X)$ induce isomorphisms on cohomology. Therefore, there exists a $\nu \in \Omega^{k-1}(U)$ such that $\omega = d\nu$. Moreover, $i^*\omega = 0 = di^*\nu$, so $i^*\nu$ is closed. Thus if we replace ν by $\nu - \pi^*i^*\nu$ we get a form whose exterior derivative is ω and whose restriction to X is zero.

Corollary 11.1.1 *The cohomology groups of the complex $\Omega(U, X)$ are zero in all dimensions.*

Let's now prove the theorem:

1. **Surjectivity.**   Given $\omega \in \Omega^k(M, X)$ with $d\omega = 0$ we can find a $\nu \in \Omega^{k-1}(U, X)$ with $\omega = d\nu$ on U. Let $\rho \in C^\infty(U)_c$ be a compactly supported C^∞ function which is one on a neighborhood of X. Then $\omega - d(\rho\nu)$ is in $\Omega^k(M - X)_c$.

2. **Injectivity.**   Given $\omega \in \Omega^k(M - X)_c$ and $\nu \in \Omega^{k-1}(M, X)$ with $\omega = d\nu$, there exists a tubular neighborhood, U, of X on which ω is zero, and hence, on which ν is closed. Thus there exists an $\alpha \in \omega^{k-2}(U, X)$ for which $\nu = d\alpha$ on U. Thus $\nu_1 = \nu - d\rho\alpha$ is in $\Omega(M - X)_c$ and $\omega = d\nu_1$.

□

Since $\Omega(M, X)$ is the kernel of the restriction $i^* : \Omega(M) \to \Omega(X)$, one has, by definition, a short exact sequence of complexes:

$$0 \to \Omega(M, X) \to \Omega(M) \to \Omega(X) \to 0 \tag{11.1}$$

and hence a long exact sequence in cohomology

$$\cdots \to H^k(\Omega(M, X)) \to H^k(M) \to H^k(X) \to H^{k+1}(\Omega(M, X)) \to \cdots$$

and hence, Theorem 11.1.1, a long exact sequence

$$\cdots \to H^k(M - X)_c \to H^k(M) \to H^k(X) \to H^{k+1}(M - X) \to \cdots$$

Suppose now that M and X are G-manifolds. Then (11.1) is a short exact sequence of G^* modules, and hence by Theorem 4.8.1 one gets a short exact sequence of equivariant de Rham complexes

$$0 \to \Omega_G(M, X) \to \Omega_G(M) \to \Omega_G(X) \to 0 \tag{11.2}$$

One also gets an inclusion map

$$\Omega_G(M - X)_c \to \Omega_G(M, X)$$

which by Theorem 6.7.1 and Theorem 11.1.1 induces an isomorphism on cohomology. Thus there is a long exact sequence

$$\cdots \to H_G^k(M - X)_c \to H_G^k(M) \to H_G^k(X) \to H_G^{k+1}(M - X)_c \to \cdots \tag{11.3}$$

11.2   Mayer-Vietoris

Another result in equivariant de Rham theory which we will need below is the equivariant version of the standard Mayer-Victoris theorem. Let M be a G-manifold and U_1 and U_2 G-invariant open subsets of M. From the short exact sequences

$$0 \to \Omega(U_1 \cup U_2) \to \Omega(U_1) \oplus \Omega(U_2) \to \Omega(U_1 \cap U_2) \to 0 \qquad (11.4)$$

and

$$0 \to \Omega(U_1 \cap U_2)_c \to \Omega(U_1)_c \oplus \Omega(U_2)_c \to \Omega(U_1 \cup U_2)_c \to 0 \qquad (11.5)$$

one gets the standard Mayer-Vietoris sequences in cohomology and in co-homology with compact supports. However, both these sequences are exact sequences of G^* modules, so by Theorem 4.8.1 one also gets long exact sequences in equivariant cohomology and in equivariant cohomology with compact supports:

$$\cdots \to H_G^{k-1}(U_1 \cap U_2) \to H_G^k(U_1 \cup U_2) \to H_G^k(U_1) \oplus H_G^k(U_2) \to \cdots \quad (11.6)$$

and

$$\cdots \to H_G^k(U_1)_c \oplus H_G^k(U_2)_c \to H_G^k(U_1 \cup U_2)_c \to H_G^{k+1}(U_1 \cap U_2) \to \cdots . \quad (11.7)$$

11.3 $S(g^*)$-Modules

Here we will review some standard facts about modules over commutative rings. (The material below can be found in any standard text on commutative algebra, for instance [AM].)

Let A be a finitely generated $S(g^*)$-module and let I_A be the annihilator ideal of A:

$$I_A = \{f \in S(g^*),\, fA = 0\}$$

The **support** of A is the algebraic variety in $g \otimes \mathbf{C}$ associated with this ideal, i.e.,

$$\operatorname{supp} A = \{x \in g \otimes \mathbf{C},\, f(x) = 0 \text{ for all } f \in I_A\} .$$

(Here we are identifying $S(g^*)$ with the ring of polynomial functions on $g \otimes \mathbf{C}$.) If $f \in I_A$ then by definition, $f = 0$ on $\operatorname{supp} A$, and conversely, if $f = 0$ on $\operatorname{supp} A$, then some power f^N lies in I_A (Hilbert's Nullstellensatz). The following is an easy exercise:

Lemma 11.3.1 *If* $A \to B \to C$ *is an exact sequence of* $S(g^*)$-*modules,* $\operatorname{supp} B \subseteq \operatorname{supp} A \cup \operatorname{supp} C$.

If A is a *free* $S(g^*)$ module the only f annihilating A is zero; so for a free module, supp $A = g \otimes \mathbf{C}$. Given an arbitrary finitely generated $S(g^*)$ module A an element $a \in A$ is defined to be a **torsion** element if $fa = 0$ for some $f \neq 0$. The set of torsion elements is a submodule of A, and A is called a **torsion** module if this submodule is A itself, i.e., if every element is a torsion element. It is clear from the definition that A is a torsion module iff supp A is a proper subset of $g \otimes \mathbf{C}$.

In the examples we will be considering in the next section, A will be a *graded* $S(g^*)$-module. Hence I_A will be a *graded* ideal, i.e., it will be generated by *homogeneous* polynomials $f_1, \ldots, f_k$. Thus supp A will be defined by the equations, $f_1 = \cdots = f_k = 0$, and hence will be a conic subvariety of $g \otimes \mathbf{C}$: if $x \in \operatorname{supp} A$ and $\lambda \in \mathbf{C}$, $\lambda x \in \operatorname{supp} A$.

11.4   The Abstract Localization Theorem

The manifolds we will be considering below will be G-manifolds, M, with "finite topology" i.e., with both $\dim H(M) < \infty$ and $\dim H(M)_c < \infty$. In this case $H_G(M)$ and $H_G(M)_c$ are, by Theorem 6.6.1, finitely generated $S(g^*)$-modules, and hence the abstract results which we described in the previous section can be applied to them. The following lemma is a simple but useful criterion for bounding the supports of these modules.

Lemma 11.4.1 *Let K be a closed subgroup of G and $\phi : M \to G/K$ a G-equivariant map. Then* supp $H_G(M)$ *and* supp $H_G(M)_c$ *are contained in* $k \otimes \mathbf{C}$.

Proof. From the maps

$$M \xrightarrow{\phi} G/K \to \mathrm{pt.}$$

one gets the inclusions of rings

$$S(g^*) = H_G(\mathrm{pt.}) \to H_G(G/K) \to H_G(M).$$

However, by Equation (4.29), $H_G(G/K) = S(k^*)^K$; so, as an $S(g^*)$-module, $H_G(M)$ is effectively an $S(k^*)$-module. That is, the ideal of functions, $f \in S(g^*)$, with $f = 0$ on $k \otimes \mathbf{C}$, annihilates it. Thus, supp $H_G(M) \subseteq k \otimes \mathbf{C}$. As for $H_G(M)_c$, it is naturally a module over the ring, $H_G(M)$; and is a module over $S(g^*)$ by the morphism of rings, $S(g^*) \to H_G(M)$. Thus it too is effectively an $S(k^*)$ module; and, as above, supp $H_G(M)_c \subseteq k \otimes \mathbf{C}$. $\square$

One situation to which this lemma applies is the following. Let p be a point in M with isotropy group K, let X be the orbit of G through p and let U be a G invariant tubular neighborhood of X. Being a tubular neighborhood, it has a G-invariant projection onto X. Therefore, identifying X with G/K we get a G equivariant map, $\phi : U \to G/K$. Moreover, for any G invariant open subset, U', of U we can restrict ϕ to U' and regard it as a G-invariant map of U' onto G/K. Thus we have proved

Lemma 11.4.2 *There exists a G-invariant neighborhood, U, of p with the property that for every G-invariant neighborhood W of p, $\operatorname{supp} H_G(U \cap W) \subseteq k \otimes \mathbf{C}$.*

By combining this with a Mayer-Vietoris argument we will prove the following key theorem.

Theorem 11.4.1 *Let M be a compact G-manifold and X a closed G-invariant submanifold. Then the supports of the modules $H_G(M-X)$ and $H_G(M-X)_c$, are contained in the set*

$$\bigcup_K k \otimes \mathbf{C} \tag{11.8}$$

the union being over subgroups K which occur as isotropy groups of points of $M - X$.

Proof. Let U be a G-invariant tubular neighborhood of X. It suffices to prove that the assertion above is true with $H_G(M-X)$ and $H_G(M-X)_c$ replaced by $H_G(M-\overline{U})$ and $H_G(M-\overline{U})_c$. Since $M-U$ is compact, one can find G invariant open sets, U_i, $i = 1, \ldots, N$ covering $M-\overline{U}$ and equivariant maps, $\phi_i : U_i \to G/K_i$, each K_i being an isotropy group of a point in $M-X$. Let $V_r = U_1 \cup \cdots \cup U_{r-1}$. We will prove by induction that the supports of $H_G(V_r)$ and $H_G(V_r)_c$ are contained in the set (11.8). By (11.6) and (11.7) there are exact sequences of $S(g^*)$ modules

$$H_G(U_r \cap V_r) \to H_G(V_{r+1}) \to H_G(U_r) \oplus H_G(V_r)$$

and

$$H_G(U_r)_c \oplus H_G(V_r)_c \to H_G(V_{r+1})_c \to H_G(U_r \cap V_r)_c$$

and for these sequences the end terms are supported in (11.8) by induction. Hence by Lemma 11.3.1 the middle terms are as well.
□

We will now describe some consequences of this theorem:

Theorem 11.4.2 *Let $i : X \to M$ be inclusion. Then the kernel and the cokernel of the map*

$$i^* : H_G(M) \to H_G(X)$$

are supported in the set (11.8).

Proof. From the exact sequence

$$H_G(M-X)_c \to H_G(M) \to H_G(X) \to H_G(M-X)_c$$

one observes that $\ker i^*$ is a quotient module of the module $H_G(M-X)_c$ and $\operatorname{coker} i^*$ is a submodule of $H_G(M-X)_c$. Thus the supports of both these modules are contained in (11.8).
□

Since M is compact there are just a finite number of distinct closed subgroups of G which can occur as isotropy groups. Let

$$\{K_i\}, \quad i = 1, \dots, \ell \tag{11.9}$$

be a list of these groups and let K be one of the groups on this list. Letting $X = M^K$ in Theorem 11.4.2 we get a result which will play a pivotal role in our proof of the Chang-Skjelbred theorem in the next section:

Theorem 11.4.3 *The kernel and cokernel of the restriction map*

$$i^* : H_G(M) \to H_G(M^K)$$

have supports in the set

$$\bigcup k_i \otimes \mathbf{C}, \quad k_i \not\supset k. \tag{11.10}$$

Let's consider in particular the case $K = G$. In this case

$$H_G(M^G) = H(M^G) \otimes S(g^*)$$

i.e., $H_G(M^G)$ is a *free* $S(g^*)$-module and hence if a is a torsion element in $H_G(M)$ it must get mapped by i^* into zero. On the other hand, by Theorem 11.4.3 the kernel of i^* is a torsion module so we conclude

Theorem 11.4.4 *(The "abstract" localization theorem)*
The kernel of the mapping

$$i^* : H_G(M) \to H_G(M^G) \tag{11.11}$$

is the module of torsion elements in $H_G(M)$.

We pointed out in Chapter 6 that there is a spectral sequence which computes the equivariant cohomology of M and whose E_1 term is $H(M) \otimes S(g^*)$. Following Goresky, Kottwitz and MacPherson [GKM] we will say that M is **equivariantly formal** if this spectral sequence collapses, in which case $H_G(M)$ is isomorphic as an $S(g^*)$-module to $H(M) \otimes S(g^*)$ and, in particular, is a *free* $S(g^*)$ module. Thus, as a corollary of Theorem 11.4.4 we get

Theorem 11.4.5 *If M is equivariantly formal the map (11.11) is injective.*

Let's come back again to the restriction map, $i^* : H_G(M) \to H_G(M^K)$, K being one of the groups, (11.9). The cokernel of i^* is supported in the set (11.10) so for every polynomial which vanishes on this set, some power of it annihilates coker i^*. In particular since the summands, $k_i \otimes \mathbf{C}$, in (11.10) do not contain k, one can find a weight, $\alpha_i \in g^*$, of G which vanishes on $k_i \otimes \mathbf{C}$ but not on k. By taking the product of these weights and then some power of these products we obtain:

Theorem 11.4.6 *There exists a monomial $p = \alpha_1 \cdots \alpha_N$, each α_i being a weight of G which does not vanish on k, such that p is in the annihilator of coker i^*. In particular for every $e \in H_G(M^K)$, pe is in the image of $i^* : H_G(M) \to H_G(M^K)$.*

11.5   The Chang-Skjelbred Theorem

Let us assume that M is equivariantly formal i.e., that $H_G(M)$ is a *free* $S(g^*)$ module. Then, as we've just shown, the map

$$i^* : H_G(M) \to H_G(M^G)$$

embeds $H_G(M)$ as a submodule of $H_G(M^G)$. Our goal in this section is to prove the following theorem of Chang-Skjelbred [CS] (see also Hsiang [Hs]).

Theorem 11.5.1 *The image of i^* is the set*

$$\bigcap_H i_H^* H_G(M^H) \tag{11.12}$$

the intersection being over all codimension-one subtori H of G and i_H being the inclusion of M^G into M^H.

Remark 11.5.1 *It is clear that the intersection (11.12) is a finite intersection only involving the codimension-one subtori, H, occurring on the list (11.9).*

Proof. The image of i^* is obviously contained in (11.12) so it suffices to prove containment in the other direction: i.e., to prove that (11.12) is contained in the image of i^*. Since i^* embeds $H_G(M)$ into $H_G(M^G)$ we can regard $H_G(M)$ as being a submodule of $H_G(M^G)$. Let $e_1, \ldots, e_K$ be a basis of $H_G(M)$ as a free module over $S(g^*)$. By Theorem 11.4.6 there exists a monomial $p = \alpha_1 \cdots \alpha_N$, the α_i's being non-zero weights of G such that for every element, e, of $H_G(M^G)$, pe is in $H_G(M)$. Hence pe can be written uniquely in the form

$$f_1 e_1 + \cdots + f_K e_K$$

with $f_i \in S(g^*)$. Dividing by p

$$e = \frac{f_1}{p} e_1 + \cdots + \frac{f_K}{p} e_K .$$

If f_i and p have a common factor we can eliminate it to write

$$e = \frac{g_1}{p_1} e_1 + \cdots + \frac{g_K}{p_K} e_K$$

with g_i and p_i relatively prime. Hence we have proved

Lemma 11.5.1 *Every element of $H_G(M^G)$ can be written uniquely in the form*

$$e = \frac{f_1}{p_1} e_1 + \cdots + \frac{f_K}{p_K} e_K \tag{11.13}$$

with $f_i \in S(g^*)$ and p_i a product of a subset of the weights, $\alpha_1, \ldots, \alpha_N$, and p_i and f_i containing no common factors.

Let us now suppose that e is in the image of $H_G(M^H)$. Then by theorem 11.4.6 there exist weights, $\{\beta_i\}$, $i = 1, \ldots, r$ of G such that $(\beta_j)|h \neq 0$ and $\beta_1 \cdots \beta_r \, e = f_1 e_1 + \cdots + f_K e_K$. We can assume without loss of generality that the β_j's are contained among the α_i's; and hence, in the "reduced representation" (11.13) of e, the p_i's occurring in the denominators are products of the β_j's. Thus we have proved

Theorem 11.5.2 *If the element (11.13) of $H_G(M^G)$ is contained in the image of $H_G(M^H)$ none of the p_i's vanish on h.*

Suppose now that this element is contained in the intersection (11.12). Let $p_i = \alpha_{i_1} \cdots \alpha_{i_s}$ and assume $s > 0$ i.e., $p_i \neq 1$. Let h be the set $\{x \in g, \alpha_{i_1}(x) = 0\}$. Since α_{i_1} is a weight, h, is the Lie algebra of a subtorus, H, of codimension-one, contradicting Theorem 11.5.2. Thus, for all i, $p_i = 1$, and $e = f_1 e_1 + \cdots f_K e_K$. Hence e is in $H_G(M)$.

11.6   Some Consequences of Equivariant Formality

The assumption "$H_G(M)$ is a free $S(g^*)$-module" has the following important consequence.

Theorem 11.6.1 *For every closed subgroup H of G each of the connected components of M^H contains a G-fixed point.*

Proof. Let X be a connected component of M^H. We will assume that $X \cap M^G = \emptyset$ and derive a contradiction. The fact that this intersection is empty implies that the G-equivariant Thom class $\tau_G(X)$ is sent to 0 by the restriction map $H_G(M) \to H_G(M^G)$. But $H_G(M)$ is a free $S(g^*)$-module, so it has no torsion. Therefore $\tau_G(X) = 0$. But under the map $H_G(M) \to H_H(M)$ induced by inclusion, $\tau_G(X)$ is sent into $\tau_H(X)$ which must therefore be 0. But then the Euler class of X in M is $e_H(N) = i^*\tau_H(X) = 0$ which contradicts Theorem 8.1. $\square$

11.7   Two Dimensional G-Manifolds

There is a standard action of $SO(3)$ on S^2 and $\mathbf{RP}^2$ and an action of T^2 on itself, and it is well known that these are basically the *only* actions of a compact connected Lie group on a two dimensional manifold. From this one easily deduces the following.

Theorem 11.7.1 *Let G be an n-dimensional torus acting non-trivially on an oriented, compact, connected two-manifold M. If M^G is non-empty, this action has the following properties:*

1. M^G is finite.

2. The set of elements of G which act trivially on M is a closed codimension-one subgroup, H.

3. M is diffeomorphic to S^2. Moreover, this diffeomorphism conjugates the action of G/H on M into the standard action of S^1 on S^2 given by rotation about the z axis.

For the sake of completeness we will sketch a proof of this theorem: Equip M with a G-invariant metric, and let p be a fixed point. The exponential map

$$\exp : T_p \to M \tag{11.14}$$

is surjective and intertwines the linear isotropy action of G on T_p with the action of G on M. Therefore, if an element of G acts trivially on T_p it also acts trivially on M. However, there exists a weight, α, of G and a linear isomorphism, $T_p \to \mathbf{C}$, conjugating the linear isotropy representation with the linear action of G on $\mathbf{C}$ given by

$$(\exp x)z = e^{i\alpha(x)}z, \qquad x \in g. \tag{11.15}$$

(See, for instance, Section 10.9.) Thus the subgroup H of G with the Lie algebra

$$\{x \in g, \, \alpha(x) = 0\}$$

acts trivially on T_p (and hence acts trivially on M). Moreover, (11.14) maps a neighborhood of 0 diffeomorphically onto a neighborhood of p, and by (11.15) $T_p^G = \{0\}$. Hence p is an isolated fixed point. Thus we have proved the first two of the assertions above.

To prove the third assertion let v be the infinitesimal generator of the action of the circle group G/H on M. By Assertion 2 the zeroes of this vector field are isolated, and from (11.15) it is easy to compute the index of v at p and show that it is one. Thus the Euler characteristic of M

$$\sum \mathrm{ind}_p v, \qquad p \in M^G$$

is just the cardinality of M^G. Hence M has positive Euler characteristic and has to be diffeomorphic to S^2.

Coming back to the invariant Riemann metric on M, by the Korn-Lichtenstein theorem (cf. [C]) this metric is conformally equivalent to the standard "round" metric on S^2 and hence G/H acts on M as a one dimensional subgroup of the group of conformal transformations $SL(2, \mathbf{C})$ of S^2. However, the connected compact one-dimensional subgroups of $SL(2, \mathbf{C})$ are all conjugate to each other; so, up to conjugacy, the action of G/H on M is the standard action of S^1 on S^2 given by rotation about the z-axis.

We will now use the results above to compute the equivariant cohomology ring, $H_G(M)$. Since H is a closed subgroup of G one has an inclusion map, $h \to g$ and hence a restriction map

$$r : S(g^*) \to S(h^*). \tag{11.16}$$

Theorem 11.7.2 *The equivariant cohomology ring $H_G(M)$ is the subring of $S(g^*) \oplus S(g^*)$ consisting of all pairs*

$$(f_1, f_2) \in S(g^*) \oplus S(g^*) \tag{11.17}$$

satisfying

$$r f_1 = r f_2. \tag{11.18}$$

Proof. Since the ordinary cohomology of M is zero in odd dimensions, M is equivariantly formal by Theorem 6.5.3; so the map, $H_G(M) \to H_G(M^G)$, is injective, and embeds $H_G(M)$ into $H_G(M^G)$ as a subring. However, M^G consists of two points, p_1 and p_2; so $H_G(M^G) = S(g^*) \oplus S(g^*)$, with one copy of $S(g^*)$ for each p_r. Moreover the embedding of $H_G(M)$ into $H_G(M^G)$ is given explicitly by

$$c \in H_G(M) \to (i_1^* c, i_2^* c) \in S(g^*) \oplus S(g^*), \tag{11.19}$$

i_r being the inclusion map, $i_r : \{p_r\} \to M$. Now note that since H acts trivially on M

$$H_H^*(M) = H^*(M) \otimes S(h^*),$$

so the natural mapping of $H_G^*(M)$ into $H_H^*(M)$ maps the element (11.19) into an element of the form

$$c' = 1 \otimes g_0 + \sum_{i=1}^{\ell} a_i \otimes g_i$$

with $a_i \in H^i(M)$, $i > 0$, and $g_0, \ldots, g_\ell$ in $S(h^*)$. Thus

$$i_1^* c' = i_2^* c' = g_0$$

giving one the compatibility condition (11.18). Thus we have proved that $H_G(M)$ is contained in the subring of $S(g^*) \oplus S(g^*)$ defined by (11.17) – (11.18). To prove that it is equal to this ring we note that

$$\begin{aligned} \dim H_G^{2k}(M) &= \dim \left(H^0(M) \otimes S^k(g^*) \oplus H^2(M) \otimes S^{k-1}(g^*) \right) \\ &= \dim S^k(g^*) + \dim S^{k-1}(g^*) \end{aligned}$$

by Theorem 6.5.1; so it suffices to check that this dimension is the same as the dimension of the $2k$-th component of the ring (11.17)–(11.18), viz.

$$2 \dim S^k(g^*) - \dim S^k(h^*)$$

i.e., to check that

$$\dim S^k(g^*) = \dim S^k(h^*) + \dim S^{k-1}(g^*).$$

This, however, follows from the fact that the restriction map, (11.16), is onto and that its kernel is $\alpha \cdot S^{k-1}(g^*)$, α being an element of $g^* - 0$ which vanishes on h. □

11.8 A Theorem of Goresky-Kottwitz-MacPherson

Let M be a compact G-manifold having the following three properties

a) $H_G(M)$ is a free $S(g^*)$-module.

b) M^G is finite.

c) For every $p \in M^G$ the weights

$$\{\alpha_{i,p}\}, \quad i = 1, \dots, d \tag{11.20}$$

of the isotropy representation of G on T_p are pairwise linearly independent: i.e., for $i \neq j$, $\alpha_{i,p}$ is not a linear multiple of $\alpha_{j,p}$.

The role of properties a) and b) is clear. We will clarify the role of property c) by proving:

Theorem 11.8.1 *Given properties a) and b) property c) is equivalent to: For every codimension one subtorus H of G, $\dim M^H \leq 2$.*

Proof. Let X be a connected component of M^H of dimension greater than zero. By Theorem 11.6.1, X contains a G-fixed point, p. Moreover,

$$T_p X = (T_p M)^H.$$

Therefore, since H is of codimension one, its Lie algebra is equal to

$$\{x \in g, \alpha_{i,p}(x) = 0\} \tag{11.21}$$

$\alpha_{i,p}$ being one of the weights on the list (11.20), and hence $T_p X$ is the one dimensional (complex) subspace of $T_p M$ associated with this weight.   □

Remark 11.8.1 *It is clear from this proof that $\dim M^H = 2$ iff the Lie algebra h of H is the algebra (11.21) for some i and p. Hence there are only a finite number of subtori*

$$\{H_i\}, \quad i = 1, \dots, N, \tag{11.22}$$

with the property that $\dim M^{H_k} = 2$, and if H is not one of the groups on the list, $M^H = M^G$.

Moreover, if H *is* one of these exceptional subtori, the connected components $\Sigma_{i,j}$ of M^{H_i} are two-spheres, and each of these two-spheres intersects M^G in exactly two points (a "north pole" and a "south pole"). For i fixed, the $\Sigma_{i,j}$'s cannot intersect each other; however, for different i's, they can intersect at points of M^G; and their intersection properties can be described by an "intersection graph" Γ whose edges are the $\Sigma_{i,j}$'s and whose vertices are the points of M^G. (Two vertices, p and q, of Γ are joined by an edge, Σ, if $\Sigma \cap M^G = \{p, q\}$.)

Moreover, for each Σ there is a unique H_i on the list (11.22) for which

$$\Sigma \subseteq M^{H_i}, \tag{11.23}$$

so the edges of Γ are *labeled* by the H_i's on this list.

Since M^G is finite

$$H_G(M^G) = H^0(M^G) \otimes S(g^*) = \text{Maps}(M^G, S(g^*))$$

and hence

$$H_G(M^G) = \text{Maps}(V_\Gamma, S(g^*)) \tag{11.24}$$

where V_Γ is the set of vertices of Γ.

Theorem 11.8.2 *[GKM] An element, p, of the ring*

$$\text{Maps}\ (V_\Gamma, S(g^*))$$

is in the image of the embedding

$$i^* : H_G(M) \rightarrow H_G(M^G)$$

if and only if for every edge Σ of the intersection graph, Γ, it satisfies the compatibility condition

$$r_h p(v_1) = r_h p(v_2) \tag{11.25}$$

v_1 and v_2 being the vertices of Σ, h being the Lie algebra of the group (11.23), and

$$r_h : S(g^*) \rightarrow S(h^*) \tag{11.26}$$

being the restriction map.

Proof. By Theorem 11.5.1 the image of i^* is the intersection:

$$\bigcap_k (i_{H_k})^* H_G(M^{H_k})$$

and by (11.18) the image of $(i_{H_k})^*$ is the set of elements of $\text{Maps}(V_\Gamma, S(g^*))$ satisfying the compatibility condition (11.25) at the vertices of Γ labelled by H_k. $\square$

11.9   Bibliographical Notes for Chapter 11

1. The main result of this chapter: that the restriction map, $H_G(M) \to H_G(M^G)$, is injective modulo torsion, is due to Borel [Bo] and, with some refinements, to Hsiang [Hs] (see Chapter 3 of [Hs], in particular the comments at the bottom of page 39).

2. Let S be the polynomial ring, $S(g^*)$, and let $S^\#$ be the quotient field of this ring, i.e., the set of all expressions

$$\frac{f}{g}, f \in S, g \in S - 0. \qquad (11.27)$$

An equivalent form of Borel's result is the assertion:

$$H_G(M) \otimes_S S^\# = H_G(M^G) \otimes_S S^\#. \qquad (11.28)$$

3. Most of the material in § 11.4 is taken verbatim from Atiyah and Bott's article, [AB]. (In particular our Theorem 11.4.2 is Theorem 3.5 of [AB].)

4. One can regard theorem 11.4.2 as a sharpening of (11.28). In order to convert the map, $H_G(M) \to H_G(M^G)$, into an isomorphism, one doesn't have to tensor with the ring of *all* quotients (11.27). It suffices to tensor with the ring of quotients

$$\frac{f}{g}, g = \alpha_1^{N_1} \dots \alpha_r^{N_r}$$

the α_i's being weights of G for which the product, g, vanishes on the set (11.4.1).

5. We have already described, in the bibliographical notes at the end of chapter 6, some criteria for M to be equivariantly formal. Here are a few additional criteria:

(a) For *every* compact G-manifold

$$\dim \sum H^i(M^G) \leq \dim \sum H^i(M),$$

and M is equivariantly formal if and only if this inequality is an equality. (See [Hs], page 46.)

(b) M is equivariantly formal iff the canonical restriction map

$$r^* : H_G(M) \to H(M)$$

is onto i.e., iff every cohomology class, $c \in H^i(M)$ is of the form

$$r^* \tilde{c} \qquad (11.29)$$

$\widetilde{c}$ being an equivariant cohomology class.

For example suppose that every *homology* class $[X]$ is representable by a G-invariant cycle, X, and hence, by Poincaré duality, that every cohomology class, c, is the Thom class of a G-invariant cycle, X. Then one can take for the $\widetilde{c}$ in (11.29) the *equivariant* Thom class of X. (Compare with item 3 in the bibliographical notes at the end of Chapter 6.)

(c) M is equivariantly formal if it possesses a G-invariant Bott-Morse function whose critical set is M^G. Indeed if such a function exists, the unstable manifolds associated with its gradient flow (and, in the case of non-isolated fixed points, the G-invariant submanifolds of these manifolds associated with G-invariant cycles in M^G) provide a basis for the homology of M. Hence M is equivariantly formal by criterion b).

(d) In particular if M is a Hamiltonian G-manifold, it is equivariantly formal as a consequence of the fact that, for a generic $\xi \in g$, the ξ-component of the moment map is a Bott-Morse function whose critical set is M^G.

(e) If M is equivariantly formal as a G-manifold, and K is a closed subgroup of G, M is equivariantly formal as a K-manifold. (This follows from the fact that the restriction map, $H_G(M) \to H(M)$, factors through $H_K(M)$.)

6. One important example of an equivariantly formal space is the compact Lie group, G, itself. For the left action of G, $H_G(G)$ is trivial; however, for the *adjoint* action the restriction map

$$H_G(G) \longrightarrow H(G) \tag{11.30}$$

is surjective and hence by criterion b) of item 5, G is equivariantly formal. (Here is a sketch of how to prove the surjectivity of (11.30) for $G = U(n)$. For N large consider G and $U(N)$ as commuting subgroups inside $U(n + N)$; and let M be the Grassmannian

$$U(N + n)/U(N) \times U(n),$$

let E be the manifold of n-frames in $\mathbf{C}^{n+N}$

$$U(n + N)/U(N),$$

and let π be the fibration of E over M with fiber, G. If $k < N$, then for every $[\mu] \in H^k(M)$, $\pi\mu = d\nu$ for some $\nu \in \Omega^{k-1}(E)$ by Proposition 2.5.2. Let $\sigma \in M$ be the identity coset, and $i : G \to E$ the inclusion of the fiber of E above σ into E. Then $i^* d\nu = i^*\pi^*\mu = 0$; so $di^*\nu = 0$; i.e., $i^*\nu$ is closed, and hence represents a cohomology class, $[i^*\nu] \in H^{k-1}(G)$. By a theorem of Koszul, these "transgressive" cohomology classes generate $H^*(G)$. (See [Kol].)

Now notice that G acts on $U(n + N)$ by its adjoint action, and that this action induces an action of G on E which is intertwined by π with the natural left action of G on M. The latter leaves fixed the point, σ; therefore the transgression construction which we outlined above can be carried out *equivariantly*. Combining this observation with Koszul's result, the question of whether (11.30) is surjective can be reduced to the question of whether the map, $H_G(M) \longrightarrow H(M)$ is surjective. However, M is a Hamiltonian G-space; so this follows from Kirwan's theorem. (See item 4 in the bibliographical notes following Chapter 6.) Notice by the way that for $i = 3$, we've already proved the surjectivity of the map, $H_G^i(G) \longrightarrow H^i(G)$ by exhibiting an explicit element of $H_G^i(G)$ whose image in $H^i(G)$ is the generator of $H^i(G)$. (See § 9.8.1.) An analogue of this "Alekseev-Malkin-Meinrenken form" in all dimensions has recently been constructed by Meinrenken and Woodward based on a construction of Jeffrey.)

7. An explicit reference for the Chang-Skjelbred theorem is: Chang, T. and Skjelbred, T., Ann. Math. 100, 307-321 (1974) page 313, lemma 2.3. Our treatment of this result in §11.5 is based on some unpublished notes of M. Brion and M. Vergne: "Sur le theorème de localization en cohomologie équivariant". (Brion and Vergne show, by the way that the assumption we have been making throughout this section, "M compact", can be replaced by the much weaker assumption: "there exists a G-invariant embedding of M into a Euclidean space on which G acts linearly".)

8. For another proof of the Chang-Skjelbred theorem within the setting of equivariant de Rham theory, using Morse theoretic techniques, see the paper of Tolman and Weitsman [TW]. Tolman and Weitsman prove that if M possesses a G-invariant Bott-Morse function whose critical set is M^G, then the "Chang-Skjelbred generators" of $H_G(M)$ can be explicitly represented by equivariant Thom classes associated with the unstable manifolds of the gradient flow of this function. (Compare with item 4d) above.)

9. The theorem of Goresky-Kottwitz-MacPherson is actually stronger than the theorem we attribute to them in § 11.8. In particular, in their paper "equivariant cohomology" means equivariant cohomology with coefficients in *sheaves*, and the result that we attribute to them is Section 11.8 is a specialization of their result to the constant sheaf **C**.

10. Let M be a compact Hamiltonian G-space with symplectic form, ω, and moment map, Φ. In addition suppose that M^G is finite and

$\Phi : M^G \longrightarrow g^*$ injective. In [GW] Ginzburg and Weinstein prove the following interesting result concerning "deformations" of the data (ω, Φ).

Theorem 11.9.1 *The data (ω, Φ) are determined up to deformation by the image of $\Phi : M^G \longrightarrow g^*$ (this image being, by the hypothesis above, a finite set of points in g^*.) More explicitly let (ω_c, Φ_c), $c \in \mathbf{R}$, be a family of G-invariant symplectic forms and moment maps depending smoothly on c with $(\omega_0, \Phi_0) = (\omega, \Phi)$. Suppose*

$$\Phi_c(M^G) = \Phi(M^G). \tag{11.31}$$

Then for c small (ω_c, Φ_c) is equivariantly diffeomorphic to (ω, Φ).

Here is a sketch of their proof: Let $\widetilde{\omega}_c$ be the equivariant symplectic form, $\omega_c - \Phi_c$. (See § 9.1.) Since M is equivariantly formal (by item 4, criterion d)) the restriction map, $H_G(M) \to H_G(M^G)$ is injective and hence by (11.31)

$$[\widetilde{\omega}] = [\widetilde{\omega}_c].$$

Thus, by the "Moser trick" (see, for instance [GS]) there is an equivariant diffeomorphism

$$f_c : M \to M$$

depending smoothly on c and equal to the identity for $c = 0$ such that

$$f_c^* \omega_c = \omega$$

and

$$\Phi_c \circ f = \Phi.$$

Appendix

Notions d'algèbre différentielle; application aux groupes de Lie et aux variétés où opère un groupe de Lie

Colloque de Topologie, C.B.R.M., Bruxelles 15–27 (1950)

1. Algèbres graduées

Soit A une algèbre (associative) sur un anneau commutatif K ayant un élément unité. Une structure graduée est définie par la donnée de sous-espaces vectoriels A^p ($p = 0$, 1, ...) tels que l'espace vectoriel A soit somme directe des A^p; un élément de A^p est dit « homogène de degré p ». On suppose de plus que le produit d'un élément de A^p et d'un élément de A^q est un élément de A^{p+q}.

On note $a \rightarrow \bar{a}$ l'automorphisme de A qui, à un élément $a \in A^p$, associe l'élément $(-1)^p a$.

Un endomorphisme θ de la structure vectorielle de A est dit *dc degré r* s'il applique A^p dans A^{p+r} pour chaque p. Parmi les endomorphismes, nous distinguerons les catégories suivantes :

1. On appelle *dérivation* tout endomorphisme θ de A, de degré *pair*, qui, vis-à-vis de la multiplication dans A, jouit de la propriété

$$\theta(ab) = (\theta a)b + a(\theta b) . \qquad (1)$$

2. On appelle *antidérivation* tout endomorphisme δ de A, de degré *impair*, qui jouit de la propriété

$$\delta(ab) = (\delta a)b + \bar{a}(\delta b) . \qquad (2)$$

Si en outre δ est de degré $+1$ et si $\delta\delta = 0$, δ s'appelle une *différentielle*; on définit alors, classiquement, l'*algèbre de cohomologie* H(A) de A, relativement à δ. C'est une algèbre graduée.

Une dérivation (resp. antidérivation) est nulle sur l'élément unité de A, s'il existe.

Si δ est une antidérivation, $\delta\delta$ est une dérivation; si δ_1 et

δ_2 sont des antidérivations, $\delta_1\delta_2 + \delta_2\delta_1$ est une dérivation. Définissons le *crochet* $[\theta_1,\theta_2]$ de deux endomorphismes θ_1 et θ_2, comme d'habitude, par la formule

$$[\theta_1,\ \theta_2] = \theta_1\theta_2 - \theta_2\theta_1\ .$$

Alors le crochet de deux dérivations est une dérivation; le crochet d'une dérivation et d'une antidérivation est une antidérivation.

Une dérivation, ou une antidérivation, est déterminée quand elle est connue sur les sous-espaces A^0 et A^1, pourvu que l'algèbre A soit engendrée (au sens multiplicatif) par ses éléments de degré 0 et 1. Dans certains cas, on peut se donner arbitrairement les valeurs d'une dérivation (ou d'une antidérivation) sur A^1, en lui donnant la valeur 0 sur A^0 : par exemple, lorsque A est l'*algèbre extérieure* d'un module M (sur K) dont les éléments sont de degré un ([1]).

Exemple. — Soit $\mathfrak{a}$ un module sur K, et soit A l'algèbre extérieure du dual $\mathfrak{a}'$ de ce module. Chaque élément x de $\mathfrak{a}$ définit un endomorphisme $i(x)$ de l'algèbre A, de degré -1, appelé « produit intérieur » par x : c'est l'unique *antidérivation*, nulle sur $A^0 = K$, qui, sur $A^1 = \mathfrak{a}'$, est égale au « produit scalaire » définissant la dualité entre $\mathfrak{a}$ et $\mathfrak{a}'$:

$$i(x) \cdot x' = \langle x, x' \rangle \text{ pour } x \in \mathfrak{a} \text{ et } x' \in A^1\ ;$$

on a alors

$$i(x) \cdot (x_1' \wedge x_2' \wedge \cdots \wedge x_p')$$
$$= \sum_{1 \leqslant k \leqslant p} (-1)^{k+1} \langle x, x_k' \rangle x_1' \wedge \cdots \wedge \widehat{x_k'} \wedge \cdots \wedge x_p' \quad (3)$$

(le signe $\wedge$ signifiant que le terme situé au-dessous doit être supprimé). L'opérateur $i(x)$ est de *carré nul* : car $i(x)i(x)$ est une dérivation, évidemment nulle sur A^0 et A^1, donc nulle partout.

Produit tensoriel d'algèbres graduées. — Soient A et B deux algèbres graduées. Sur le produit tensoriel $A \otimes B$ de leurs espaces vectoriels, considérons la loi multiplicative définie par $(a \otimes b) \cdot (a' \otimes b') = (-1)^{p'q} (aa') \otimes (bb')$, si b est de degré q et a' de degré p'. Définissons sur $C = A \otimes B$ une structure graduée, en appelant C^r le sous-espace de C, somme directe des $A^p \otimes B^q$ tels que $p + q = r$. Alors C est munie d'une structure d'*algèbre graduée*; cette algèbre graduée s'appelle le produit tensoriel des algèbres graduées A et B.

Le cas le plus intéressant est celui où A et B ont un élément unité, les sous-algèbres A^0 et B^0 étant isomorphes à l'an-

---

([1]) Pour ce qui concerne les algèbres extérieures en général, voir BOURBAKI, *Algèbre*, chap. III.

neau de base K. Dans ce cas, on identifiera toujours A à une sous-algèbre de $A \otimes B$, par l'application biunivoque $a \longrightarrow a \otimes 1$ de A dans $A \otimes B$ (on note 1 l'élément unité); de même, on identifiera B à une sous-algèbre de $A \otimes B$. En outre, désignons par B^+ la somme directe des B^q pour $q \geqslant 1$; $A \otimes B$ est somme directe de la sous-algèbre $A \otimes B^0$ (identifiée à A) et de l'idéal $A \otimes B^+$. Cette décomposition directe définit un projecteur de $A \otimes B$ sur $A \otimes B^0$, donc une application linéaire de $A \otimes B$ sur A; cette application est compatible avec les structures multiplicatives; nous l'appellerons la *projection canonique* de $A \otimes B$ sur A. Elle identifie A à l'algèbre quotient de $A \otimes B$ par l'idéal $A \otimes B^+$. On définit de même la projection canonique de $A \otimes B$ sur B.

Plaçons-nous toujours dans l'hypothèse où $A^0 = B^0 = K$. Soit donnée une application linéaire θ_1 de A dans $C = A \otimes B$, de degré pair, satisfaisant à la condition (1), et une application linéaire θ_2 de B dans C, de même degré, satisfaisant aussi à (1). Il existe alors une *dérivation* θ de l'algèbre $A \otimes B$, et une seule, qui se réduise à θ_1 sur A et à θ_2 sur B; elle est définie par

$$\theta(a \otimes b) = \theta_1(a) \cdot b + a \cdot \theta_2(b) \qquad (4)$$

(le signe $\cdot$ désignant la multiplication dans $A \otimes B$).

De même, étant données une application linéaire δ_1 de A dans C, de degré impair, satisfaisant à (2), et une application linéaire δ_2 de B dans C, de même degré, satisfaisant aussi à (2), il existe une *antidérivation* δ de l'algèbre $A \otimes B$, et une seule, qui se réduise à δ_1 sur A et à δ_2 sur B; elle est définie par

$$\delta(a \otimes b) = \delta_1(a) \cdot b + \bar{a} \cdot \delta_2(b). \qquad (5)$$

2. Variétés différentiables

Pour simplifier l'exposition, on se bornera aux variétés indéfiniment différentiables. Les *champs de vecteurs tangents* que l'on considérera seront toujours supposés indéfiniment différentiables; de même, les formes différentielles extérieures, de tous degrés, seront supposées à coefficients indéfiniment différentiables.

Les champs de vecteurs tangents constituent un module T sur l'anneau K des fonctions numériques (indéfiniment différentiables). Le module dual T' est le module des formes différentielles *de degré un*. L'algèbre extérieure $A(T')$ du module T' est l'algèbre des formes différentielles de tous degrés (les fonctions, éléments de K, ne sont autres que les formes différentielles de degré 0). La différentiation extérieure, notée d, est une « différentielle » sur $A(T')$, au sens du § 1. Chaque élément x de T définit, outre le *produit intérieur* $i(x)$ (qui opère sur $A(T')$ et est de degré -1), une « transformation infinitésimale » $\theta(x)$ qui opère aussi bien sur T que sur T' et $A(T')$; sur $A(T')$, c'est une *dérivation* de degré 0, qui est entièrement caractérisée par les deux conditions suivantes :

$$\theta(x)d = d\theta(x) \text{ (c'est-à-dire : } \theta(x) \text{ commute avec } d) \text{ ;} \quad (6)$$

$$\theta(x) \cdot f = i(x) \cdot df \text{ pour toute } \textit{fonction } f \in A^\circ(T') \text{ .} \quad (7)$$

Si x et y sont deux champs de vecteurs tangents, le champ de vecteurs $\theta(x) \cdot y$ se note $[x, y]$; cette notation se justifie parce que

$$\theta([x, y]) = \theta(x)\theta(y) - \theta(y)\theta(x) \text{ .} \quad (I)$$

En outre, sur l'algèbre différentielle $A(T')$, on a les relations

$$\theta(x)i(y) = i(y)\theta(x) + i([x, y]) \text{ ,} \quad (II)$$

$$\theta(x) = i(x)d + di(x) \quad (III)$$

(formule qui, compte tenu de $dd = 0$, entraîne la relation n° 6).

3. Groupes de Lie

Soit G un groupe de Lie connexe. Les champs de vecteurs tangents, invariants par les translations à gauche, forment un *espace vectoriel* $a(G)$ sur le corps réel; cet espace est en dualité avec l'espace vectoriel $a'(G)$ des formes différentielles de degré un, invariantes à gauche. L'algèbre extérieure $A(G)$ de $a'(G)$ est l'algèbre (sur le corps réel) des formes différentielles de tous degrés, invariantes à gauche. Les éléments de degré 0 (fonctions constantes) s'identifient aux scalaires (multiples de l'unité). L'algèbre différentielle $A(G)$ a une algèbre de cohomologie qui, lorsque G est *compact*, s'identifie à l'algèbre de cohomologie (réelle) de l'espace G.

Chaque élément x de $a(G)$ définit un groupe à un paramètre d'automorphismes de G, qui ne sont autres que les *translations à droite* par un sous-groupe à un paramètre de G. La transformation infinitésimale $\theta(x)$ de ce groupe opère dans $a(G)$; donc $a(G)$ est stable pour le crochet $[x, y]$. L'espace $a(G)$, muni de la structure définie par ce crochet, est l'*algèbre de Lie* du groupe G. En outre, les $\theta(x)$ opèrent dans $A(G)$, ainsi que les produits intérieurs $i(x)$. Sur l'algèbre différentielle $A(G)$, les opérateurs d, $i(x)$ et $\theta(x)$ satisfont aux relations (I), (II) et (III) du paragraphe précédent.

Ici, on peut expliciter l'opérateur différentiel d de $A(G)$: désignons par $e(x')$ la multiplication (à gauche) par un élément $x' \in A^1(G)$ dans l'algèbre $A(G)$; alors, en prenant dans

$\mathfrak{a}(G)$ et dans son dual $\mathfrak{a}'(G) = A^1(G)$ deux bases duales (x_k) et (x_k'), on a

$$d = \frac{1}{2} \sum_k e\,(x_k')\,\theta\,(x_k)\,. \qquad \text{(IV)}$$

Cette formule a été donnée par Koszul dans sa thèse ([1]). Appliquée aux éléments de degré un de $A(G)$, elle donne les « équations de Maurer-Cartan ».

4. Espace fibré principal

C'est une variété $\mathscr{E}$, que nous supposerons indéfiniment différentiable, et où un groupe de Lie connexe G opère de manière que :

1° L'application $(P, s) \rightarrow P \cdot s$ de $\mathscr{E} \times G$ dans $\mathscr{E}$ soit indéfiniment différentiable; et $(P \cdot s) \cdot t = P \cdot (st)$ (ce qu'on exprime en disant que G opère « à droite »);

2° G soit simplement transitif dans chaque classe d'équivalence (fibre);

3° L'espace $\mathscr{B}$ (« espace de base ») quotient de $\mathscr{E}$ par la relation d'équivalence définie par G, soit une variété indéfiniment différentiable;

4° Chaque point de $\mathscr{B}$ possède un voisinage ouvert $\mathscr{U}$ tel que l'image réciproque de $\mathscr{U}$ dans $\mathscr{E}$ soit isomorphe (comme variété indéfiniment différentiable) au produit $\mathscr{U} \times G$, la transformation définie par un élément s de G étant alors (u, g) $\rightarrow (u, gs)$.

On notera E l'algèbre des formes différentielles (à coefficients indéfiniment différentiables) de l'espace $\mathscr{E}$, munie de sa graduation et de l'opérateur d de différentiation extérieure. Tout vecteur x de l'algèbre de Lie $\mathfrak{a}(G)$ définit un *champ de vecteurs tangents* à $\mathscr{E}$: en effet, chaque fibre de $\mathscr{E}$ s'identifie à G, d'une manière bien déterminée à une translation à gauche près du groupe G; donc un champ invariant à gauche, sur G, se transporte sur chaque fibre d'une seule manière. Ainsi, chaque élément x de $\mathfrak{a}(G)$ définit, dans l'algèbre E, un produit intérieur $i(x)$ et une transformation infinitésimale $\theta(x)$; et les relations (I), (II) et (III) du § 2 sont satisfaites.

D'ailleurs la transformation infinitésimale $\theta(x)$ n'est autre que celle du sous-groupe à un paramètre de G (groupe d'opérateurs à droite dans $\mathscr{E}$) défini par l'élément x de l'algèbre de Lie $\mathfrak{a}(G)$.

([1]) *Bull. Soc. math. de France*, 1950, pp. 65-127; voir formule (3.4), p. 74.

L'algèbre B des formes différentielles de l'espace de base $\mathcal{B}$ s'identifie à une *sous-algèbre* de E, stable pour d, à savoir la sous-algèbre des éléments annulés par tous les opérateurs $i(x)$ et $\theta(x)$ relatifs aux éléments x de $\mathfrak{a}(G)$.

D'une manière générale, soit E une algèbre différentielle graduée où opèrent des antidérivations $i(x)$ (de degré —1 et de carré nul) et des dérivations $\theta(x)$ (de degré 0) correspondant aux éléments x d'une algèbre de Lie $\mathfrak{a}(G)$, de manière à satisfaire à (I), (II) et (III). Nous dirons, pour abréger, que G *opère dans l'algèbre* E. Cela étant, nous appellerons *éléments basiques* de E les éléments annulés par tous les $i(x)$ et les $\theta(x)$; ils forment une sous-algèbre graduée B, *stable pour d* [en vertu de (III)]. On appelle éléments *invariants* de E les éléments annulés par les $\theta(x)$; ils forment une sous-algèbre *stable pour d*, que nous noterons I_E.

Dans certains cas, l'homomorphisme canonique $H(I_E) \longrightarrow H(E)$ des algèbres de cohomologie de I_E et de E est un *isomorphisme* de la première *sur* la seconde. Il en est ainsi notamment dans les cas suivants : 1) E est de dimension finie, et l'algèbre de Lie $\mathfrak{a}(G)$ est *réductive* (i.e. : composée directe d'une algèbre abélienne et d'une algèbre semi-simple); 2) E est l'algèbre des formes différentielles d'un espace fibré principal $\mathcal{E}$ dont le groupe G est *compact*.

On a un homomorphisme canonique $H(B) \longrightarrow H(I_E)$. Un problème important consiste à chercher des relations plus précises entre les algèbres de cohomologie $H(I_E)$ et $H(B)$; dans le cas 2) ci-dessus, ce sont respectivement les algèbres de cohomologie de l'espace fibré $\mathcal{E}$ et de son espace de base $\mathcal{B}$.

5. CONNEXION INFINITÉSIMALE DANS UN ESPACE FIBRÉ PRINCIPAL

Une connexion infinitésimale est définie par la donnée, en chaque point P de l'espace fibré $\mathcal{E}$, d'un *projecteur* φ_P de l'espace tangent à $\mathcal{E}$ au point P, sur le sous-espace des vecteurs tangents à la fibre au point P, de manière que :

1° φ_P soit fonction indéfiniment différentiable du point P;

2° Les projecteurs φ_P relatifs aux points d'une même fibre se transforment les uns dans les autres par les opérations du groupe G.

On peut prouver l'*existence* de telles connexions infinitésimales [1]. De plus, la donnée d'une connexion infinitésimale dans $\mathcal{E}$ revient à la donnée d'une application linéaire f du dual $A^1(G)$ de l'algèbre de Lie $\mathfrak{a}(G)$, dans le sous-espace E^1

---

[1] Voir la conférence de Ch. Ehresmann à ce Colloque.

des éléments de degré un de l'algèbre E, application qui satisfasse aux deux conditions suivantes :

$$i(x) \cdot f(x') = i(x) \cdot x'$$
(scalaire de E, c'est-à-dire fonction constante sur $\mathscr{E}$),
$$\theta(x) \cdot f(x') = f(\theta(x) \cdot x') \qquad \qquad (8)$$

pour tout $x \in \mathfrak{a}(G)$ et tout $x' \in A^1(G)$.

Supposons qu'on ait un autre espace fibré principal $\mathscr{E}'$ de même groupe G, et un G-*homomorphisme* de $\mathscr{E}'$ dans $\mathscr{E}$ (c'est-à-dire une application indéfiniment différentiable de $\mathscr{E}'$ dans $\mathscr{E}$, compatible avec les opérations de G dans $\mathscr{E}'$ et $\mathscr{E}$). Un tel homomorphisme définit d'une manière évidente l'image réciproque d'une connexion infinitésimale sur $\mathscr{E}$: c'est une connexion infinitésimale sur $\mathscr{E}'$. On vérifie aisément que l'application f' de $A^1(G)$ dans E'^1 définie par cette dernière est composée de l'application f et de l'homomorphisme de E dans E' défini par l'application de l'espace $\mathscr{E}'$ dans l'espace $\mathscr{E}$.

Ce qui précède conduit à la notion abstraite de « connexion algébrique » dans une algèbre différentielle E (avec élément-unité) dans laquelle opère un groupe G (au sens de la fin du § 4) : ce sera une application linéaire de $A^1(G)$ dans E^1 qui satisfasse aux conditions (8).

Soit alors f une telle connexion algébrique. Supposons en outre que l'algèbre E satisfasse à la loi d'anticommutation $vu = (-1)^{pq}uv$ pour u de degré p et v de degré q. Alors on peut prolonger f, d'une seule manière, en un *homomorphisme (multiplicatif) de l'algèbre* $A(G)$ *dans l'algèbre* E, qui transforme l'élément unité de $A(G)$ dans l'élément unité de E. Notons encore f ce prolongement. On a alors, pour tout élément $a \in A(G)$,

$$i(x) \cdot f(a) = f(i(x) \cdot a)$$
$$\theta(x) \cdot f(a) = f(\theta(x) \cdot a) \, . \qquad (8')$$

Autrement dit, f est *compatible* avec les opérateurs $i(x)$ et $\theta(x)$, qui opèrent dans $A(G)$ et dans E.

Mais, si $x' \in A^1(G)$, *on n'a pas*, en général, $d(f(x')) = f(dx')$; autrement dit, f n'est pas compatible avec les différentielles de $A(G)$ et de E. L'application $x' \longrightarrow d(f(x')) - f(dx')$ de $A^1(G)$ dans E^2 est ce qu'on appelle le *tenseur de courbure* de la connexion.

L'élément $d(f(x')) - f(dx')$ n'est pas, en général, un élément basique de E; toutefois il est *annulé par tous les produits intérieurs* $i(x)$. Démonstration :

$$i(x)d \cdot (f(x')) = \theta(x) \cdot f(x') - d \cdot (i(x) \cdot f(x')) = f(\theta(x) \cdot x')$$

d'après (8) et, d'après (8'),

$$i(x) \cdot f(dx') = f(i(x) \cdot dx') = f(\theta(x) \cdot x') - f(di(x) \cdot x')$$
$$= f(\theta(x) \cdot x') \,.$$

6. L'algèbre de Weil d'une algèbre de Lie

Les considérations précédentes ont conduit André Weil (dans un travail non publié) à associer à l'algèbre de Lie $\alpha(G)$ une autre algèbre différentielle, dont $A(G)$ est un quotient, et que nous allons définir maintenant.

Désignons par $S(G)$ l'*algèbre symétrique* du dual $\alpha'(G)$ de $\alpha(G)$. Si on prend une base (x_k') dans $\alpha'(G)$, $S(G)$ s'identifie à l'algèbre des *polynômes* par rapport aux lettres x_k' (commutant deux à deux). $S(G)$ s'identifie aussi canoniquement à l'algèbre des formes multilinéaires *symétriques* sur l'espace vectoriel $\alpha(G)$.

On distinguera l'espace $\alpha'(G)$ comme sous-espace $A^1(G)$ de $A(G)$, et comme sous-espace $S^1(G)$ de $S(G)$. On a un isomorphisme canonique h de $A^1(G)$ sur $S^1(G)$. On notera souvent $\tilde{x}'$ l'élément $h(x')$, pour $x' \in A^1(G)$.

Si on a une connexion algébrique f de $A^1(G)$ dans une algèbre E comme ci-dessus, et qu'on prolonge f en un homomorphisme de $A(G)$ dans E, on est amené à définir une application linéaire $\tilde{f}$ de $S^1(G)$ dans E^2, en posant

$$\tilde{f}(\tilde{x}') = d(f(x')) - f(dx')$$

Pour que $\tilde{f}$ conserve les degrés, on convient que les éléments de $S^1(G)$ sont de *degré 2*. Ceci conduit à graduer $S(G)$ en convenant que les éléments de $S^p(G)$ (formes p-linéaires symétriques sur $\alpha(G)$) sont de degré $2p$. L'application $\tilde{f}$ se prolonge alors en un homomorphisme multiplicatif, de degré 0, de l'algèbre (commutative) $S(G)$ dans l'algèbre E.

On notera encore $\tilde{f}$ l'homomorphisme prolongé.

L'*algèbre de Weil* de l'algèbre de Lie $\alpha(G)$ sera, par définition, l'algèbre graduée

$$W(G) = A(G) \otimes S(G) \,,$$

produit tensoriel des algèbres graduées $A(G)$ et $S(G)$ (cf. § 1).

Les homomorphismes $f : A(G) \rightarrow E$, et $\tilde{f} : S(G) \rightarrow E$, définissent un homomorphisme $\bar{f}$ de l'algèbre $W(G)$ dans l'algèbre E, par la formule

$$\bar{f}(a \otimes s) = f(a) \otimes \tilde{f}(s)$$

L'homomorphisme (multiplicatif) $\bar{f}$ est de degré 0.

On va définir, sur $W(G)$, *d'une manière indépendante de*

l'algèbre E *et de la connexion* f, des opérateurs $i(x)$, $\theta(x)$ et une différentielle δ, de telle manière que, pour toute connexion f dans une algèbre différentielle graduée É dans laquelle opère G (avec la loi d'anticommutation $vu = (-1)^{pq}uv$), l'homomorphisme $\bar{f}$ défini par f soit *compatible* avec les opérateurs $i(x)$, $\theta(x)$ et les opérateurs différentiels δ (de W(G)) et d (de E).

Définition de $i(x)$: $i(x)$ est déjà défini sous la sous-algèbre A(G) de W(G) = A(G) $\otimes$ S(G). Sur S(G), convenons que l'opérateur $i(x)$ est nul. Sur W(G), $i(x)$ sera l'unique antidérivation qui prolonge $i(x)$ sur A(G) et 0 sur S(G); on a $i(x)i(x) = 0$, car $i(x)i(x)$ est une dérivation nulle sur A(G) et sur S(G). Cela posé, on a, pour tout élément $w \in$ W(G),

$$i(x) \cdot \bar{f}(w) = \bar{f}(i(x) \cdot w) \quad \text{(compatibilité de } \bar{f} \text{ avec } i(x)),$$

parce qu'il en est ainsi lorsque w est dans A(G) (relation (8')) et lorsque w est dans S^1(G) (les deux membres étant alors nuls).

Définition de $\theta(x)$: $\theta(x)$ est déjà défini sur A(G). On va le définir sur S^1(G), puis on le prolongera en une dérivation (de degré 0) sur W(G) = A(G) $\otimes$ S(G). Or soit $\tilde{x}' \in S^1$(G), donc $\tilde{x}' = h(x')$; on pose $\theta(x) \cdot \tilde{x}' = h(\theta(x) \cdot x')$; ceci définit $\theta(x)$ sur S^1(G).

$\theta(x)$ étant alors prolongé à W(G), on a bien, pour tout $w \in$ W(G),

$$\theta(x) \cdot \bar{f}(w) = \bar{f}(\theta(x) \cdot w)$$

(compatibilité de $\bar{f}$ avec $\theta(x)$), parce qu'il en est ainsi lorsque w est dans A(G) (relations (8')), et lorsque w est dans S^1(G); en effet

$$\theta(x) \cdot \tilde{f}(\tilde{x}') = \tilde{f}(\theta(x) \cdot \tilde{x}') \quad \text{(vérification immédiate, grâce à (8'))}.$$

Il y a intérêt à décomposer l'opérateur $\theta(x)$ sur W(G) en la somme de deux opérateurs partiels :

$$\theta(x) = \theta_A(x) + \theta_s(x) ,$$

où $\theta_A(x)$ est égal à $\theta(x)$ sur A(G) et nul sur S(G), et $\theta_s(x)$ est égal à $\theta(x)$ sur S(G) et nul sur A(G).

Définition de l'opérateur différentiel δ *de* W(G). — La relation

$$d(f(x')) = f(dx') + \tilde{f}(\tilde{x}')$$

conduit à poser (si l'on veut que $\bar{f}$ soit compatible avec les opérateurs différentiels)

$$\delta x' = dx' + \tilde{x}' = dx' + h(x') . \tag{9}$$

De même la relation

$$i(x)d \cdot \tilde{f}(\tilde{x'}) = \theta(x) \cdot \tilde{f}(\tilde{x'}) = \tilde{f}(\theta(x) \cdot \tilde{x'})$$

conduit à poser

$$i(x) \cdot \tilde{\delta x'} = \theta(x) \cdot \tilde{x'} \, ,$$

ou, ce qui revient au même,

$$\tilde{\delta x'} = \sum_k x_k' \otimes \theta(x_k) \cdot \tilde{x'} \tag{10}$$

$((x_k)$ et (x_k') étant deux bases duales).

L'opérateur δ étant ainsi défini sur $A^1(G)$ et $S^1(G)$, il se prolonge d'une seule manière en une antidérivation (notée encore δ) de $W(G) = A(G) \otimes S(G)$. Son degré est $+1$. Les formules (9) et (10) permettent d'ailleurs d'expliciter δ sur $W(G)$ tout entier :

$$\delta = d_A + d_S + h \, , \tag{11}$$

où h prolonge l'application (déjà notée h) de $A^1(G)$ dans $S^1(G)$ en une antidérivation de $W(G)$, nulle sur $S(G)$:

$$h = \sum_k i(x_k) e(\tilde{x_k'}) \tag{12}$$

$(e(\tilde{x'})$ désigne la multiplication par l'élément $\tilde{x'} \in S^1(G))$.

Quant aux opérateurs d_A et d_S, ils sont explicités par les formules suivantes (dont la première résulte de la formule (IV) du § 3) :

$$d_A = \frac{1}{2} \sum_k e(x_k') \theta_A(x_k) \, , \tag{13}$$

$$d_S = \sum_k e(x_k') \theta_S(x_k) \, . \tag{14}$$

Les formules (11), (12), (13), (14) explicitent complètement l'opérateur δ de $W(G)$.

La relation analogue à (III) (§ 2) :

$$\theta(x) = i(x)\delta + \delta i(x)$$

a lieu sur $W(G)$. Elle résulte des relations suivantes :

$$i(x)h + hi(x) = 0 \, , \tag{15}$$
$$\theta_A(x) = i(x)d_A + d_A i(x) \, , \tag{16}$$
$$\theta_S(x) = i(x)d_S + d_S i(x) \, . \tag{17}$$

On vérifie que δ est une différentielle : $\delta\delta = 0$. Il suffit de vérifier que la dérivation $\delta\delta$ est nulle sur $A^1(G)$ et $S^1(G)$. Ainsi, $W(G)$ est une algèbre différentielle graduée, munie d'opérateurs $i(x)$ et $\theta(x)$ satisfaisant aux conditions (I), (II), (III) du § 2.

Chaque fois qu'on a une connexion algébrique dans une algèbre différentielle graduée E où opère G [avec la loi d'anti-commutation $vu = (-1)^{pq}uv$], on obtient un homomorphisme $\bar{f}$ de l'algèbre de Weil $W(G)$ dans l'algèbre E, compatible avec les graduations, les opérateurs $i(x)$ et $\theta(x)$, et enfin *compatible avec les opérateurs différentiels*. Seuls, ce dernier point reste à vérifier; or la relation

$$\bar{f}(\delta w) = d(\bar{f}(w))$$

a lieu si w est dans $A^1(G)$ ou dans $S^1(G)$, d'après la manière même dont δ a été défini (cf. relations (9) et (10)); il en résulte qu'elle a lieu pour tout $w \in A(G) \otimes S(G)$. C. Q. F. D.

Cas particulier. — Prenons pour E l'algèbre $A(G)$, f étant l'application identique de $A^1(G)$ dans $A^1(G)$. Alors $\bar{f}$ n'est autre que la *projection canonique* (cf. fin du § 1) de $W(G) = A(G) \otimes S(G)$ sur $A(G)$; cette projection canonique est compatible avec les opérateurs différentiels δ (de $W(G)$) et $d = d_A$ (de $A(G)$). Donc l'algèbre différentielle $A(G)$ s'identifie à un quotient de l'algèbre différentielle $W(G)$.

Remarque. — Si le groupe G est *abélien*, les opérateurs $\theta(x)$ sont nuls; alors la différentielle δ de $W(G)$ se réduit à l'antidérivation h définie par la formule (12).

7. Classes caractéristiques (réelles) d'un espace fibré principal

Soit $\mathscr{E}$ un espace fibré principal de groupe G, et soit E l'algèbre des formes différentielles de l'espace $\mathscr{E}$. Il existe une *connexion* f, d'où un homomorphisme $\bar{f}$ de l'algèbre de Weil $W(G)$ dans E.

Les éléments basiques de $W(G)$ sont transformés par $\bar{f}$ dans des éléments basiques de E, c'est-à-dire des éléments de l'algèbre B des formes différentielles de l'espace de base $\mathscr{B}$. Or les éléments basiques de $W(G)$ ne sont autres que les éléments *invariants* de $S(G)$. Nous noterons $I_s(G)$ la sous-algèbre de ces éléments invariants; elle s'identifie à l'algèbre des formes multilinéaires symétriques sur $\mathfrak{a}(G)$, invariantes par le groupe adjoint; on n'oubliera pas que les éléments de $I^p_s(G)$ sont de *degré 2 p*. Dans la seconde conférence, nous étudierons la structure de l'algèbre $I_s(G)$.

La formule (14) montre ceci : pour qu'un élément de $S(G)$ soit un *cocycle* de $W(G)$ (c'est-à-dire pour que le δ de cet élément soit nul), il faut et il suffit qu'il soit *invariant*.

Revenant à l'homomorphisme $\bar{f}$, il applique les éléments de $I_s(G)$, qui sont des cocyles de $W(G)$, dans des cocyles de B, c'est-à-dire des *cocyles de l'espace de base*. On les appelle les *cocyles caractéristiques de la connexion*; ils sont de *degrés pairs*. Ils forment une sous-algèbre du centre de B, appelée la sous-algèbre caractéristique de la connexion.

Si on a un G-homomorphisme d'un espace fibré principal $\mathcal{E}'$ (de groupe G) dans l'espace fibré $\mathcal{E}$, et qu'on envisage la connexion f' qu'il définit (image réciproque de la connexion f), l'homomorphisme $\bar{f}'$ de $W(G)$ dans E' est évidemment composé de $\bar{f}$ et de l'homomorphisme de E dans E' (défini par l'application de $\mathcal{E}'$ dans $\mathcal{E}$). Donc l'homomorphisme $I_s(G) \longrightarrow B'$ est composé de $I_s(G) \longrightarrow B$ et de l'homomorphisme $B \longrightarrow B'$.

Passons maintenant des cocyles de B à leurs classes de cohomologie, éléments de l'algèbre de cohomologie $H(B)$. La connexion f définit un *homomorphisme de $I_s(G)$ dans $H(B)$*, *qui applique* $I^p_s(G)$ *dans* $H^{2p}(B)$. Cet homomorphisme, introduit par A. Weil, joue un rôle fondamental; on verra (2° conférence) qu'il est *indépendant du choix de la connexion*. C'est donc un *invariant de la structure fibrée* de l'espace $\mathcal{E}$. L'image de $I_s(G)$ par cet homomorphisme est une sous-algèbre de l'algèbre de cohomologie $H(B)$ de l'espace de base, appelée la sous-algèbre caractéristique de la structure fibrée; ses éléments sont les *classes caractéristiques* de la structure fibrée; elles sont de *degrés pairs*.

Si on a un G-homomorphisme d'un espace fibré principal $\mathcal{E}'$, de même groupe G, dans l'espace $\mathcal{E}$, l'homomorphisme $H(B) \longrightarrow H(B')$ qu'il définit applique la sous-algèbre caractéristique de $H(B)$ *sur* la sous-algèbre caractéristique de $H(B')$.

8. L'ALGÈBRE DE WEIL COMME ALGÈBRE UNIVERSELLE

On sait ([1]) que si un espace fibré principal $\mathcal{E}$, de groupe G, est tel que ses groupes d'homotopie $\pi_i(\mathcal{E})$ soient nuls pour $0 \leqslant i \leqslant N$ ($\pi_0(\mathcal{E}) = 0$ signifiant que $\mathcal{E}$ est connexe), alors, pour tout espace fibré principal $\mathcal{E}'$, de groupe G, dont la base $\mathcal{B}'$ est un espace de dimension $\leqslant N$, il existe un G-homomorphisme de $\mathcal{E}'$ dans $\mathcal{E}$; d'où un homomorphisme $E \longrightarrow E'$ et un homomorphisme $B \longrightarrow B'$. De plus, deux quelconques de ces G-homomorphismes définissent des applications de $\mathcal{B}'$ dans $\mathcal{B}$ qui sont *homotopes*, et par suite l'homomorphisme

---

([1]) Voir par exemple STEENROD, *Annals of Math.*, 45, 1944, pp. 294-311, pour le cas où G est le groupe orthogonal.

$H(B) \longrightarrow H(B')$ est univoquement déterminé par la structure fibrée de $\mathcal{E}'$. Un tel espace $\mathcal{E}$ est dit *classifiant pour la dimension* N.

Par exemple, si G est le groupe orthogonal, on connaît des espaces classifiants pour des dimensions arbitrairement grandes (mais un même espace n'est pas classifiant pour toutes les dimensions). Leurs bases sont des grassmanniennes (réelles).

Revenant à l'algèbre de Weil $W(G)$, on voit qu'elle se comporte, du point de vue homologique, comme une algèbre universelle pour les espaces fibrés de groupe G, c'est-à-dire comme une algèbre de cochaînes d'un espace fibré qui serait classifiant pour tous les espaces fibrés de groupe G, quelle que soit la dimension de leur espace de base. L'algèbre $I_s(G)$ joue le rôle de l'algèbre des cochaînes de l'espace de base d'un tel espace fibré universel, avec la particularité que les éléments de $I_s(G)$ sont tous des cocycles. L'homomorphisme de $W(G)$ dans E', défini par une connexion dans l'algèbre E' des cochaînes de l'espace $\mathcal{E}'$, joue le rôle que jouait l'homomorphisme $E \longrightarrow E'$ défini par un G-homomorphisme d'un espace classifiant $\mathcal{E}$ dans l'espace $\mathcal{E}'$; l'homomorphisme $I_s(G) \longrightarrow B'$ joue le rôle que jouait l'homomorphisme $B \longrightarrow B'$; enfin, l'homomorphisme (unique) $I_s(G) \longrightarrow H(B')$ joue le rôle que jouait l'homomorphisme (unique) $H(B) \longrightarrow H(B')$.

En fait, on verra, dans la deuxième conférence (§ 7), que si, G étant *compact* (connexe), l'espace $\mathcal{E}$ est classifiant pour la dimension N, alors $H^m(B)$ est nul pour les *m impairs* $\leqslant$ N, et l'homomorphisme canonique $I_s(G) \longrightarrow H(B)$ applique *biunivoquement* $I''_s(G)$ *sur* $H^{2p}(B)$ pour $2p \leqslant$ N. Ceci donnera une preuve, *a priori*, du fait que les espaces de cohomologie des bases de deux espaces classifiants pour la dimension N sont isomorphes pour tous les degrés $\leqslant$ N.

La transgression dans un groupe de Lie et dans un espace fibré principal

Colloque de Topologie, C.B.R.M., Bruxelles 57–71 (1950)

Les notations de la première conférence (^{1}) sont conservées. En particulier, $I_s(G)$ continue à désigner l'algèbre des éléments invariants de $S(G)$, c'est-à-dire des éléments basiques de l'algèbre de Weil $W(G) = A(G) \otimes S(G)$. De plus, nous introduirons les notations suivantes : $I_A(G)$ pour l'algèbre des éléments invariants de $A(G)$, et $I_w(G)$ pour l'algèbre des éléments invariants de $W(G)$. Ce sont des algèbres différentielles graduées (l'opérateur différentiel étant induit, pour $I_A(G)$, par celui de l'algèbre ambiante $A(G)$, et, pour $I_w(G)$, par celui de l'algèbre ambiante $I_w(G)$).

En fait, l'opérateur différentiel de $I_A(G)$ *est nul*, en vertu de la formule (IV) de la première conférence (§ 3).

On notera $H_A(G)$ l'algèbre de cohomologie de $A(G)$, qui, lorsque G est un groupe compact (connexe), s'identifie à l'algèbre de cohomologie réelle de l'espace compact G. Puisque les éléments de $I_A(G)$ sont des cocycles, on a un homomorphisme canonique $I_A(G) \longrightarrow H_A(G)$. Il est bien connu que, lorsque l'algèbre de Lie $\mathfrak{a}(G)$ est *réductive* (c'est-à-dire composée directe d'une algèbre abélienne et d'une algèbre semi-simple), l'homomorphisme $I_A(G) \longrightarrow H_A(G)$ est une application *biunivoque de* $I_A(G)$ *sur* $H_A(G)$. On en trouvera une démonstration dans la thèse de Koszul (^{2}). Ceci vaut notamment lorsque G est un groupe compact.

1. La cohomologie de l'algèbre de Weil

En relation avec le caractère universel de l'algèbre de Weil (1$^\text{re}$ conférence, § 8), on a le théorème suivant :

(^{1}) Page 15 de ce Recueil.
(^{2}) *Bull. Soc. Math. de France*, 1950, pp. 65-127; voir le théorème 9.2 du chapitre IV.

Théorème 1. — *L'algèbre de cohomologie de* $W(G)$ *est triviale :* $H^m(W(G))$ *est nul pour tout entier* $m \geqslant 1$ (pour $m = 0$, $H^0(W(G))$ s'identifie évidemment au corps des scalaires). *De même, l'algèbre de cohomologie de la sous-algèbre* $I_W(G)$ *est triviale.*

Ce théorème vaut sans aucune hypothèse restrictive sur l'algèbre de Lie $\mathfrak{a}(G)$. Il se démontre comme suit : soit k l'antidérivation de $W(G)$, de degré -1, nulle sur $A(G)$, et définie sur $S^1(G)$ par $k(\tilde{x}') = x'$ (autrement dit : l'endomorphisme composé kh est l'identité sur $A^1(G)$). L'opérateur k commute avec les transformations infinitésimales $\theta(x)$, par suite k opère dans la sous-algèbre $I_W(G)$ des éléments invariants de $W(G)$.

$\delta k + k\delta$ est une *dérivation*; elle est entièrement définie quand on la connaît sur $A^1(G)$ et sur $S^1(G)$: or elle transforme tout $x' \in A^1(G)$ en x' lui-même, et tout $\tilde{x}' \in S^1(G)$ en $\tilde{x}' - d_A x'$. Appelons *poids* d'un élément de $W(G)$ le plus grand des entiers q tels que sa composante dans $A(G) \otimes S^q(G)$ ne soit pas nulle (le poids étant, par définition, -1 si l'élément considéré est nul). Soit alors u un élément homogène de degré m ($m \geqslant 1$) de $W(G)$; $\delta k u + k\delta u$ est homogène de degré m. Soit $q \geqslant 0$ le poids de u (u étant supposé $\neq 0$); le poids de

$$v = u - \frac{1}{m-q}(\delta k u + \delta k u)$$

est $\leqslant q - 1$. Le processus qui fait passer de u à l'élément v de poids strictement plus petit peut être itéré, et conduira finalement à un élément nul. *Supposons que* u *soit un cocycle :* $\delta u = 0$; alors v est un cocycle homologue à u, et de proche en proche on voit que u est le cobord d'un élément de $W(G)$. Ceci montre bien que $H^m(W(G))$ est nul.

Si en outre u est un cocycle *invariant*, le processus montre que u est le cobord d'un élément invariant de $W(G)$; donc $H^m(I_W(G))$ est nul.

2. L'application canonique $I_S^p(G) \to I_A^{2p-1}(G)$.

Soit $u \in I_S^p(G)$ ($p \geqslant 1$). Puisque c'est un cocycle de degré $2p$ de l'algèbre $I_W(G)$, il existe, d'après le théorème 1, un $w \in I_W(G)$, de degré $2p - 1$, tel que $\delta w = u$. La projection canonique de $W(G)$ sur $A(G)$ transforme w en un élément w_A de $I_A(G)$, de degré $2p - 1$. Cet élément ne dépend pas du choix de w; car si $\delta w' = \delta w$, il existe un $v \in I_W(G)$ tel que $w' - w = \delta v$. Alors $w_A' - w_A = d_A v_A$, et comme $v_A \in I_A(G)$, $d_A v_A$ est nul.

En associant ainsi à chaque $u \in I_S^p(G)$ l'élément $w_A \in I_A^{2p-1}(G)$, on définit une *application linéaire canonique de* $I_S^p(G)$ *dans* $I_A^{2p-1}(G)$, pour toute valeur de l'entier $p \geqslant 1$; cette application sera notée ρ.

Les éléments de l'image de cet homomorphisme jouissent de la propriété d'être *transgressifs* dans l'algèbre $I_W(G)$. Voici ce qu'on entend par là : un élément $a \in I_A^q(G)$ est dit transgressif s'il est l'image, par la projection canonique de $W(G)$ sur $A(G)$, d'un élément $w \in I_W^q(G)$ dont le cobord δw soit dans $S(G)$, et par suite dans $I_S(G)$. Alors w s'appelle une *cochaîne de transgression* pour a; un élément transgressif a peut avoir plusieurs cochaînes de transgression.

Tout élément transgressif non nul est de degré impair : car si a transgressif est de degré pair, δw est de degré impair, et comme δw est dans $I_S(G)$ dont tous les degrés sont pairs, δw est nul. D'après le théorème 1, il existe alors un $v \in I_W(G)$ tel que $\delta v = w$; d'où $a = w_A = d_A v_A$, et comme $v_A \in I_A(G)$, cela implique $d_A v_A = 0$.

Les éléments transgressifs de $I_A(G)$ forment un sous-espace vectoriel $T_A(G)$, engendré par des éléments de degrés impairs; $T_A(G)$ est le sous-espace de $I_A(G)$, *image de l'application* ρ. Prenons une base homogène de $T_A(G)$, et, à chaque élément a de cette base, associons le cobord $\delta w \in I_S(G)$ d'une cochaîne de transgression w. On obtient une application linéaire de $T_A(G)$ dans $I_S(G)$, qu'on appellera *une transgression*. On peut encore définir une transgression comme suit : c'est une application linéaire τ de $T_A(G)$ dans $I_S(G)$, qui, suivie de l'application canonique ρ, donne l'application identique de $T_A(G)$.

3. La transgression
dans le cas d'une algèbre de Lie réductive

Rappelons d'abord le théorème de Hopf ([1]) : si $a(G)$ est réductive, l'algèbre $I_A(G)$ s'identifie à l'*algèbre extérieure* d'un sous-espace bien déterminé $P_A(G)$ de $I_A(G)$; l'espace $P_A(G)$ est engendré par des éléments homogènes de degrés impairs, appelés éléments *primitifs* de $I_A(G)$; la dimension de $P_A(G)$ est le *rang* $r(G)$ du groupe G. Voici comment on définit un élément homogène primitif : considérons l'algèbre extérieure $A_*(G)$ de l'algèbre de Lie $a(G)$ (algèbre des « chaînes » du groupe G), et la sous-algèbre $I_*(G)$ des éléments invariants de $A_*(G)$. L'hypothèse de réductivité entraîne que la dualité canonique entre $A_*(G)$ et $A(G)$ induit une dualité entre $I_*(G)$ et $I_A(G)$;

([1]) Voir thèse de Koszul, chap. IV, § 10.

cela étant, un élément homogène de $I_\Lambda^p(G)$ $(p \geqslant 1)$ est appelé *primitif* s'il est orthogonal aux éléments (de degré p) *décomposables* de $I_*(G)$ (dans une algèbre graduée quelconque, un élément homogène de degré p est décomposable s'il est somme de produits d'éléments homogènes de degrés strictement plus petits que p).

Ceci étant rappelé, revenons à l'application ρ et à la transgression :

THÉORÈME 2. — *Si l'algèbre de Lie est réductive, l'image de l'application canonique ρ est l'espace $P_\Lambda(G)$ des éléments primitifs de $I_\Lambda(G)$ (autrement dit, $P_\Lambda(G)$ est identique à l'espace $T_\Lambda(G)$ des éléments transgressifs). Le noyau de l'application ρ est formé des éléments décomposables de $I_s(G)$.*

Ce théorème a d'abord été conjecturé par A. Weil en mai 1949; le fait que tout élément primitif est transgressif a aussitôt été prouvé par Chevalley, s'inspirant d'une démonstration donnée par Koszul du théorème de transgression de sa thèse (th. 18.3). Puis H. Cartan a démontré qu'il n'y a, dans l'image de ρ, que des éléments primitifs, et que le noyau est formé exactement des éléments décomposables de $I_s(G)$.

Le théorème 2 (qu'il n'est pas question de démontrer ici) entraîne ceci : pour toute transgression $\tau : P_\Lambda(G) \rightarrow I_s(G)$, l'image de τ *engendre* (au sens multiplicatif) l'algèbre (commutative) $I_s(G)$. Une étude plus approfondie (Chevalley, Koszul; cf. la conférence de Koszul à ce Colloque) montre que les transformés, par une transgression τ, des éléments d'une *base* homogène de $P_\Lambda(G)$, sont *algébriquement indépendants* dans $I_s(G)$; par suite $I_s(G)$ a la structure d'une *algèbre de polynômes* à $r(G)$ variables ($r(G)$: rang du groupe G). D'une façon plus précise, le nombre des générateurs de poids p de l'algèbre $I_s(G)$ est égal à la dimension de l'espace des éléments primitifs de $I_\Lambda(G)$, de degré $2p - 1$.

Ce résultat relatif à la structure de l'algèbre $I_s(G)$ est le pendant du théorème de Hopf sur la structure de l'algèbre $I_\Lambda(G)$.

4. TRANSGRESSION DANS UN ESPACE FIBRÉ PRINCIPAL DE GROUPE G

L'algèbre de Lie est désormais supposée *réductive*.

Soit, avec les notations de la première conférence, E l'algèbre des formes différentielles d'un espace fibré principal $\mathcal{E}$, de groupe G (groupe de Lie connexe, tel que son algèbre de Lie $\mathfrak{a}(G)$ soit réductive). Choisissons une connexion infinitésimale dans $\mathcal{E}$; elle définit un homomorphisme $\bar{f}$ de $W(G)$ dans E, compatible avec les graduations et tous les opérateurs.

Soit alors a un élément primitif de $I_\Lambda(G)$ (cocycle invariant de la *fibre* de l'espace $\mathscr{E}$) ; choisissons une cochaîne de transgression w (comme il a été dit au § 2) ; alors $\bar{f}(w)$ est un élément de E qui « induit » le cocycle a sur chaque fibre. Sa différentielle $d\bar{f}(w) = \bar{f}(\delta w)$ est l'élément de B (algèbre des formes différentielles de l'espace de base $\mathscr{B}$) que la connexion associe à l'élément δw de $I_s(G)$. Ainsi la forme différentielle $\bar{f}(w)$ (« forme de transgression ») *a pour différentielle un cocycle de l'espace de base*.

Il est ainsi prouvé que les cocycles invariants primitifs de la fibre sont « transgressifs » dans l'espace fibré principal $\mathscr{E}$; fait qui a d'abord été mis en évidence par Koszul dans le cas particulier où $\mathscr{E}$ est l'espace d'un groupe de Lie dont G est un sous-groupe ([1]), puis a été généralisé par A. Weil en se servant de la transgression dans $W(G)$, comme il vient d'être expliqué.

Faisons choix une fois pour toutes d'une transgression τ dans $W(G)$; alors le choix d'une connexion f dans E définit une forme de transgression $\bar{f}(w)$ pour chaque cocycle invariant primitif $a \in P_\Lambda(G)$; l'application linéaire $a \rightarrow d\bar{f}(w)$ de $P_\Lambda(G)$ dans B, appelée « transgression » dans l'espace fibré, applique $P_\Lambda^{2p-1}(G)$ dans B^{2p} ; elle est composée de la transgression $\tau : P_\Lambda^{2p-1}(G) \rightarrow I_s^p(G)$, et de l'application $I_s^p(G) \rightarrow B^{2p}$ définie par la connexion (cf. première conférence, § 7).

Soit φ l'application linéaire $P_\Lambda(G) \rightarrow B$ ainsi obtenue. Sur l'algèbre graduée $I_\Lambda(G) \otimes B$, il existe une antidérivation et une seule qui, sur le sous-espace $P_\Lambda(G)$ de $I_\Lambda(G)$, soit égale à φ, et, sur B, soit égale à la différentielle d de B. Cette antidérivation Δ est de degré $+1$, et son carré est nul : c'est une différentielle. Un théorème de Chevalley ([2]) permet d'affirmer, lorsque G est un groupe *compact* (connexe), que l'algèbre de cohomologie de $I_\Lambda(G) \otimes B$, pour la différentielle Δ, s'identifie canoniquement à l'algèbre de cohomologie de la sous-algèbre I_E des éléments invariants de E. D'ailleurs $H(I_E)$ s'identifie canoniquement à l'algèbre $H(E)$, algèbre de cohomologie de l'espace fibré $\mathscr{E}$. En résumé : *la connaissance de l'homomorphisme $I_s(G) \rightarrow B$ défini par une connexion de l'espace fibré permet de définir, sur l'algèbre $I_\Lambda(G) \otimes B$, une différentielle pour laquelle l'algèbre de cohomologie s'identifie à l'algèbre de cohomologie (réelle) de l'espace fibré.* En particulier : quand on connaît l'espace de base $\mathscr{B}$, et l'homomorphisme $I_s(G) \rightarrow B$ défini par une connexion, on connaît l'algèbre de cohomologie (réelle) de l'espace fibré.

([1]) Thèse, théorème 18.3.
([2]) Voir la conférence de Koszul à ce Colloque.

5. Recherche de la cohomologie de l'espace de base

Nous nous intéressons désormais au problème inverse du précédent : il s'agit de trouver un processus qui permette, de la cohomologie H(E) de l'espace fibré, de passer à la cohomologie H(B) de l'espace de base. Pour cela, nous nous placerons dans le cadre algébrique général : E est une algèbre différentielle graduée dans laquelle opère un groupe de Lie G (dans le sens du § 4 de la première conférence); B est alors la sous-algèbre des *éléments basiques* de E.

Considérons le produit tensoriel E ⊗ W(G) (produit tensoriel d'algèbres graduées). C'est une algèbre graduée, sur laquelle nous considérons la différentielle $\bar{\delta}$ qui prolonge la différentielle d de E et la différentielle δ de W(G). De plus, les antidérivations $i(x)$ (déjà définies sur E et sur W(G)) se prolongent en antidérivations de E ⊗ W(G), que l'on notera encore $i(x)$; on définit de même les dérivations $\theta(x)$ sur E ⊗ W(G). Il est clair que les relations (I), (II) et (III) de la première conférence (où d serait remplacé par $\bar{\delta}$) sont satisfaites sur E ⊗ W(G), puisqu'elles le sont sur E et sur W(G).

Soit $\bar{B}$ la sous-algèbre des éléments basiques de E ⊗ W(G) : éléments annulés par les $i(x)$ et les $\theta(x)$. Elle est stable pour $\bar{\delta}$, et l'on peut considérer l'algèbre de cohomologie H($\bar{B}$).

Si l'on songe à l'interprétation de W(G) comme algèbre de cochaînes d'un espace fibré universel (cf. § 8 de la première conférence), les opérations précédentes admettent l'interprétation géométrique suivante : soit $\mathscr{E}'$ un espace fibré classifiant; considérons l'espace produit $\mathscr{E} \times \mathscr{E}'$, et faisons-y opérer le groupe G par la loi : $(P, P') \longrightarrow (P \cdot s, P' \cdot s)$. L'espace quotient $\mathscr{F}$ est un espace fibré de même base $\mathscr{B}$ et de fibre $\mathscr{E}'$. L'algèbre E ⊗ W(G) joue alors le rôle de l'algèbre des cochaînes de l'espace $\mathscr{E} \times \mathscr{E}'$, et $\bar{B}$ joue le rôle de l'algèbre des cochaînes de l'espace fibré $\mathscr{F}$. Or, dans un sens à préciser, la fibre $\mathscr{E}'$ est « triviale »; cela laisse supposer que la cohomologie de $\mathscr{F}$ s'identifie à la cohomologie de l'espace de base $\mathscr{B}$. En fait, nous allons voir que, sous certaines hypothèses, on peut identifier H($\bar{B}$) et H(B).

Les algèbres différentielles B et I_s(G) (l'opérateur différentiel de la seconde est d'ailleurs nul) s'identifient canoniquement à des sous-algèbres de l'algèbre différentielle $\bar{B}$. On en déduit des homomorphismes *canoniques*

$$H(B) \longrightarrow H(\bar{B}) \quad \text{et} \quad I_s(G) \longrightarrow H(\bar{B}) \tag{1}$$

Théorème 3. — *S'il existe une « connexion »* (au sens algébrique du mot) *dans* E, *l'homomorphisme* H(B) $\longrightarrow$ H($\bar{B}$) *est un isomorphisme de* H(B) *sur* H($\bar{B}$).

Ceci étant admis, le second homomorphisme (1) donne un homomorphisme $I_s(G) \longrightarrow H(B)$, et on voit facilement que c'est précisément celui que définit la connexion (première conférence, § 7). Par conséquent, l'homomorphisme défini par une connexion est *indépendant de la connexion*, comme il avait été annoncé.

Pour démontrer le théorème 3, on utilise sur

$$E \otimes W(G) = E \otimes A(G) \otimes S(G) ,$$

l'antidérivation k, de degré -1, qui est nulle sur E et A(G), et est définie, sur $S^1(G)$, par

$$k(1 \otimes 1 \otimes \tilde{x}') = 1 \otimes x' \otimes 1 - f(x') \otimes 1 \otimes 1 ,$$

f désignant l'application $A^1(G) \longrightarrow E^1$ définie par la connexion. On raisonne alors comme dans la démonstration du théorème 1, en considérant la dérivation $\overline{\delta}k + k\overline{\delta}$; la récurrence est un peu plus subtile, elle permet de montrer que tout élément de $\overline{B}$ dont la différentielle est dans B est la somme d'un élément de B et de la différentielle d'un élément de $\overline{B}$.

6. Transformation de l'algèbre différentielle $\overline{B}$

$\overline{B}$ est contenue dans la sous-algèbre de $E \otimes A(G) \otimes S(G)$ formée des éléments annulés par les produits intérieurs $i(x)$, c'est-à-dire dans le produit tensoriel $F \otimes S(G)$, F désignant la sous-algèbre de $E \otimes A(G)$ formée des éléments annulés par les $i(x)$. D'une façon précise, $\overline{B}$ s'identifie à la sous-algèbre des éléments *invariants* de $F \otimes S(G)$.

Pour interpréter F, considérons la projection canonique de $E \otimes A(G)$ sur E; elle commute avec les $\theta(x)$ et applique *biunivoquement* la sous-algèbre F *sur* E, comme on s'en assure aisément. D'où un isomorphisme canonique de F sur E, qui permet d'identifier $\overline{B}$ à la sous-algèbre C des éléments *invariants* de $E \otimes S(G)$. Reste à expliciter la différentielle Δ que l'on obtient en transportant à C la différentielle $\overline{\delta}$ de $\overline{B}$: on trouve que Δ est induite, sur C, par la différence $d - h$ des antidérivations d et h de $E \otimes S(G)$ que voici :

d se réduit, sur E, à la différentielle de E, et est nulle sur $S(G)$;

h est nulle sur $S(G)$, et est donnée par la formule

$$h = \sum_k i(x_k) e(\tilde{x_k'}) \qquad (2)$$

(où (x_k) et (x_k') sont deux bases duales de $\mathfrak{a}(G)$ et $A^1(G)$,

$e(\widetilde{x_k}')$ désignant la multiplication par $\widetilde{x_k}'$ dans l'algèbre $E \otimes S(G)$).

On notera que le carré de $d - h$ n'est pas nul en général; mais sa restriction Δ à la sous-algèbre C a un carré nul.

Reste à voir ce que deviennent les homomorphismes

$$I_s(G) \longrightarrow H(B) \quad et \quad H(B) \longrightarrow H(I_E)$$

quand on identifie $H(B)$ à $H(C)$. On voit aussitôt que le premier est défini en considérant $I_s(G)$ comme une sous-algèbre de C, tandis que le second s'obtient à partir de l'application de C sur I_E définie par la projection canonique de $E \otimes S(G)$ sur E. Résumons :

THÉORÈME 4. — *S'il existe une connexion dans E, l'algèbre de cohomologie* $H(B)$ *de la sous-algèbre B des éléments basiques de E s'identifie canoniquement à l'algèbre de cohomologie* $H(C)$ *de la sous-algèbre C des éléments invariants de* $E \otimes S(G)$, *munie de la différentielle* Δ *explicitée ci-dessus. Par cette identification, l'homomorphisme* $I_s(G) \longrightarrow H(B)$ *devient l'homomorphisme* $I_s(G) \longrightarrow H(C)$ *obtenu en considérant* $I_s(G)$ *comme sous-algèbre de C, et l'homomorphisme* $H(B) \longrightarrow H(I_E)$ *devient l'homomorphisme* $H(C) \longrightarrow H(I_E)$ *obtenu en considérant* I_E *comme algèbre quotient de C.*

Observons que si G est un groupe compact (connexe), ou si, E étant de dimension finie, $\alpha(G)$ est réductive, $H(I_E)$ s'identifie canoniquement à $H(E)$.

Remarque. — Examinons le cas où E est l'algèbre $A(G)$ elle-même, la connexion étant définie par l'application identique de $A^1(G)$ dans $A^1(G)$ (cf. première conférence, fin du § 6). Alors C est la sous-algèbre des éléments invariants de $A(G) \otimes S(G)$, c'est-à-dire la sous-algèbre $I_w(G)$; l'opérateur h est le même que celui défini par la formule (12) de la première conférence; et on constate que $\Delta = d - h$, sur $I_w(G)$, est égale à $-\delta$, δ étant la différentielle de l'algèbre de Weil $W(G)$.

7. UTILISATION DE LA THÉORIE DE HIRSCH-KOSZUL

La théorie de Hirsch ([1]), mise au point par Koszul ([2]), peut s'appliquer à l'algèbre C et conduit aux résultats suivants :

Supposons que G soit un groupe *compact* (connexe); ou que, E étant de dimension finie, l'algèbre $\alpha(G)$ soit *réductive*. Alors il existe une application linéaire φ, biunivoque (mais non déterminée de manière unique) de l'espace vectoriel gra-

---

([1]) *Comptes rendus*, 227, 1948. p. 1328.
([2]) Dans un travail non encore publié.

dué $H(E) \otimes I_s(G)$ sur un sous-espace vectoriel de l'espace vectoriel gradué C, application qui conserve les degrés et possède les propriétés suivantes :

1. Sur $I_s(G)$ (sous-algèbre de $H(E) \otimes I_s(G)$), φ se réduit à l'application identique ($I_s(G)$ étant aussi identifiée à une sous-algèbre de C);

2. φ applique chaque élément $a \otimes 1$ de $H(E) \otimes 1$ sur un élément de C dont la projection canonique (élément de I_E) est un cocycle de la classe de a;

3. L'image de $H(E) \otimes I_s(G)$ par φ est *stable pour* Δ.

Grâce à φ, on peut alors identifier $H(E) \otimes I_s(G)$ à un sous-espace vectoriel de C, ce qui définit sur $H(E) \otimes I_s(G)$ un opérateur cobord (de carré nul, obtenu en transportant Δ); et l'on montre que l'application de l'espace de cohomologie de $H(E) \otimes I_s(G)$ (relatif à cet opérateur cobord) dans $H(C)$ est *biunivoque sur*. (Par contre, comme φ n'est pas, en général, un homomorphisme multiplicatif, il n'y a pas de structure multiplicative sur $H(H(E) \otimes I_s(G))$.)

Finalement, on voit qu'il existe sur $H(E) \otimes I_s(G)$ un opérateur cobord, de degré $+1$, nul sur $I_s(G)$, et qui applique $H(E)$ dans l'idéal engendré par $I_s^+(G)$; de plus, il existe un isomorphisme de l'espace de cohomologie $H(H(E) \otimes I_s(G))$ sur $H(B)$, compatible avec les homomorphismes du diagramme

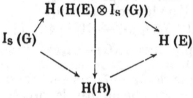

Application. — Supposons que $\mathcal{E}$ soit un espace fibré principal de groupe compact connexe G, *classifiant* pour la dimension N (cf. § 8 de la première conférence). Alors les espaces de cohomologie $H^m(E)$ sont nuls pour $1 \leqslant m \leqslant N$, et $H^0(E)$ se réduit au corps des scalaires. Dans ces conditions, l'espace de cohomologie de $H(E) \otimes I_s(G)$ s'identifie à $I_s(G)$ pour tous les degrés $m \leqslant N$. Ceci prouve que l'homomorphisme $I_s(G) \rightarrow H(B)$ est un *isomorphisme* de $I_s^p(G)$ *sur* $H^{2p}(B)$ pour $2p \leqslant N$, et que $H^m(B)$ est nul pour les valeurs *impaires* de $m \leqslant N$. C'est le résultat annoncé à la fin de la première conférence.

8. Réduction du groupe structural

Soit g un sous-groupe fermé, connexe, du groupe compact G; ou encore, dans un cadre purement algébrique, suppo-

sons que E soit de dimension finie, que l'algèbre de Lie $\alpha(G)$ et
sa sous-algèbre $\alpha(g)$ soient réductives. On notera B_G ce qui était
noté B auparavant : B_G sera la sous-algèbre de E, formée des
éléments annulés par les $i(x)$ et les $\theta(x)$ relatifs aux éléments
$x \in \alpha(G)$. De même, B_g désignera la sous-algèbre de E formée
des éléments annulés par les $i(x)$ et les $\theta(x)$ relatifs aux
$x \in \alpha(g)$. On notera C_G (resp. C_g) la sous-algèbre des éléments G-invariants de $E \otimes S(G)$ (resp. des éléments g-invariants de $E \otimes S(g)$). L'homomorphisme naturel de $S(G)$ dans
$S(g)$ définit un homomorphisme de $E \otimes S(G)$ dans $E \otimes S(g)$,
qui applique C_G dans C_g et est compatible avec les différentielles; d'où un homomorphisme $H(C_G) \rightarrow H(C_g)$, qui s'identifie à l'homomorphisme $H(B_G) \rightarrow H(B_g)$ quand on identifie
$H(C_G)$ à $H(B_G)$ et $H(C_g)$ à $H(B_g)$.

Cela étant, remontons à la définition d'un opérateur
cobord δ_G dans l'espace vectoriel $H(E) \otimes I_s(G)$, obtenu à partir
d'une application φ de $H(E) \otimes I_s(G)$ dans C_G, qui satisfasse
aux conditions 1, 2 et 3 du § 7. Cette application est déterminée quand on connaît sa restriction φ_0 au sous-espace
$H(E) \otimes 1$. Composons φ_0 avec l'application canonique de C_G
dans C_g; on obtient une application ψ_0 de $H(E) \otimes 1$ dans C_g,
qui, prolongée à $H(E) \otimes I_s(g)$, donne une application ψ satisfaisant aussi aux conditions 1, 2 et 3. D'où un opérateur cobord
δ_g dans $H(E) \otimes I_s(g)$, pour lequel l'espace de cohomologie de
$H(E) \otimes I_s(g)$ s'identifie à l'espace $H(B_g)$.

Finalement : l'application canonique

$$H(E) \otimes I_s(G) \rightarrow H(E) \otimes I_s(g)$$

(définie par l'application $I_s(G) \rightarrow I_s(g)$ déduite de $S(G) \rightarrow$
$S(g)$) associe à tout opérateur cobord δ_G de $H(E) \otimes I_s(G)$
(obtenu par le procédé du § 7) un opérateur cobord δ_g dans
$H(E) \otimes I_s(g)$, lui aussi obtenu par le procédé du § 7. Pour
définir δ_g sur $H(E) \otimes 1$, on effectue successivement δ_G puis l'application canonique de $H(E) \otimes I_s(G)$ dans $H(E) \otimes I_s(g)$.

Application. — Soit $\mathcal{E}$ un espace fibré principal dont le
groupe G est compact et connexe; soit g un sous-groupe fermé,
connexe, de G. Notons $\mathcal{B}_G$ l'espace quotient de $\mathcal{E}$ par G, et
$\mathcal{B}_g$ l'espace quotient de $\mathcal{E}$ par g. L'application canonique
$H(\mathcal{B}_G) \rightarrow H(\mathcal{B}_g)$ est donnée par

$$H(H(E) \otimes I_s(G)) \rightarrow H(H(E) \otimes I_s(g)) .$$

Lorsque g a *même rang* que G, l'application $I_s(G) \rightarrow I_s(g)$
est *biunivoque* (voir ci-dessous). Il en résulte très facilement
que *l'application* $H(\mathcal{B}_G) \rightarrow H(\mathcal{B}_g)$ *est biunivoque*. On retrouve
ainsi, au moins dans le cas des espaces fibrés indéfiniment
différentiables, un résultat de J. Leray [1].

[1] *Comptes rendus*, 25 juillet 1949.

La transgression dans un groupe de Lie 215

9. Cohomologie des espaces homogènes

Nous allons appliquer la théorie précédente à la recherche de l'algèbre de cohomologie (réelle) d'un espace homogène G/g, lorsque G est un groupe compact connexe et g un sous-groupe fermé connexe de G. Cette question a déjà fait l'objet d'importants travaux de Samelson, Leray et Koszul [2]. Nous nous placerons ici dans le cadre algébrique suivant : $a(G)$ sera une algèbre de Lie réductive, et $a(g)$ une sous-algèbre *réductive dans* $a(G)$ [3].

E sera alors l'algèbre $A(G)$, dans laquelle opèrent les $i(x)$ et les $\theta(x)$ relatifs aux $x \in a(g)$. La sous-algèbre B des « éléments basiques » de E s'identifie, lorsque G est un groupe compact connexe et g un sous-groupe fermé connexe, à l'algèbre des formes différentielles de l'espace homogène G/g, invariantes à gauche par G; et l'on sait que son algèbre de cohomologie s'identifie à l'algèbre de cohomologie $H(G/g)$ de l'espace homogène. C'est pourquoi nous noterons désormais $H(G/g)$ l'algèbre de cohomologie $H(B)$.

D'autre part, l'algèbre de cohomologie $H(G)$ de l'espace fibré G s'identifie canoniquement à la sous-algèbre $I_A(G)$ des éléments G-invariants de $A(G)$.

Le théorème 4 est applicable, parce qu'il existe une « connexion » dans $E = A(G)$: une application linéaire f de $A^1(g)$ dans $A^1(G)$, compatible avec les $i(x)$ et les $\theta(x)$ relatifs aux $x \in a(g)$. Une telle connexion est définie par un projecteur de l'algèbre de Lie $a(G)$ sur la sous-algèbre $a(g)$, projecteur qui soit compatible avec les $\theta(x)$ relatifs aux $x \in a(g)$; ou encore, par un sous-espace vectoriel de $a(G)$, supplémentaire de $a(g)$, et stable par les transformations infinitésimales de $a(g)$. L'existence d'un tel sous-espace résulte de l'hypothèse suivant laquelle $a(g)$ est réductive dans $a(G)$.

Appliquons le théorème 4 : $H(G/g)$ s'identifie canoniquement à l'algèbre de cohomologie de la sous-algèbre C_g des éléments g-invariants de l'algèbre $A(G) \otimes S(g)$, munie de l'opérateur Δ_g explicité au théorème 4. Mais ici, non seulement la théorie de Hirsch-Koszul est applicable comme au § 7, mais (cf. la conférence de Koszul) on peut astreindre l'application φ du § 7 à être compatible avec les structures *multiplicatives*. D'une façon précise, l'algèbre $H(E) \otimes I_s(G)$ envisagée au § 7 devient ici $I_A(G) \otimes I_s(g)$; et φ va être un isomorphisme de cette

---

[2] Références bibliographiques dans la thèse de Koszul; voir en outre les *Notes* de Leray aux *Comptes rendus*, t. 228, 1949, p. 1902, et t. 229, 1949, p. 280.

[3] Pour cette notion, voir thèse de Koszul, § 9.

algèbre sur une sous-algèbre de l'algèbre C_g. Pour définir φ, on remarque que $I_\Lambda(G)$ est l'algèbre extérieure du sous-espace $P_\Lambda(G)$ de ses éléments primitifs; par suite φ sera déterminé quand on le connaîtra sur le sous-espace $P_\Lambda(G) \otimes 1$ de $I_\Lambda(G) \otimes I_s(g)$. Pour définir φ sur ce sous-espace de manière à satisfaire aux conditions 1, 2 et 3 du § 7, on observe que *tout élément de* $P_\Lambda(G)$ *est transgressif dans l'algèbre* C_g (voir ci-dessous), de sorte qu'on définira φ en associant, à tout élément $a \otimes 1$ de $P_\Lambda(G) \otimes 1$, une cochaîne de transgression dans l'algèbre C_g des éléments g-invariants de $A(G) \otimes S(g)$.

Démontrons que tout élément de $P_\Lambda(G)$ est *transgressif* dans C_g, comme il a été annoncé. On va, pour cela, se servir à nouveau de la transgression dans l'algèbre de Weil de a (G) : considérons l'homomorphisme canonique λ de $A(G) \otimes S(G)$ dans $A(G) \otimes S(g)$, défini par $S(G) \longrightarrow S(g)$. Il applique la sous-algèbre $I_w(G)$ des éléments G-invariants de $A(G) \otimes S(G) = W(G)$ dans la sous-algèbre C_g des éléments g-invariants de $A(G) \otimes S(g)$; en outre, il est compatible avec les opérateurs différentiels Δ_G et Δ_g de $I_w(G)$ et C_g respectivement (cf. la remarque de la fin du § 6). Soit alors a un élément de $P_\Lambda(G)$; il est transgressif dans $I_w(G)$; si w est une cochaîne de transgression de $a \otimes 1$ dans $I_w(G)$, $\lambda(w)$ sera, dans C_g, une cochaîne de transgression pour $a \otimes 1$, puisque sa différentielle $\Delta_g \lambda(w)$ sera égale à L'image, par l'application $I_s(G) \longrightarrow I_s(g)$, de $\Delta_G(w)$, qui est un élément de $I_s(G)$.

Ceci prouve en outre que, pour obtenir une transgression $P_\Lambda(G) \longrightarrow I_s(g)$ dans l'algèbre C_g, il suffit de prendre une transgression $P_\Lambda(G) \longrightarrow I_s(G)$ (cf. § 3), et de la composer avec l'homomorphisme canonique $I_s(G) \longrightarrow I_s(g)$. Résumons les résultats obtenus :

THÉORÈME 5. — *Choisissons une transgression* $\tau : P_\Lambda(G) \longrightarrow I_s(G)$, *et composons-la avec l'homomorphisme canonique de* $I_s(G)$ *dans* $I_s(g)$. *On obtient une application linéaire de* $P_\Lambda(G)$ *dans* $I_s(g)$, *qu'on prolonge en une antidérivation de* $I_\Lambda(G) \otimes I_s(g)$, *nulle sur* $I_s(g)$. *Cette antidérivation, de degré* $+1$, *est une différentielle* δ *sur l'algèbre graduée* $I_\Lambda(G) \otimes I_s(g)$; *l'algèbre de cohomologie de* $I_\Lambda(G) \otimes I_s(g)$, *pour* δ, *est isomorphe à l'algèbre de cohomologie* $H(G/g)$ *de l'espace homogène* G/g, *par un isomorphisme compatible avec les homomorphismes du diagramme*

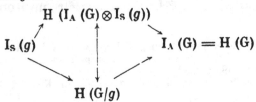

(Dans ce diagramme, l'homomorphisme

$$I_s(g) \longrightarrow H(I_\Lambda(G) \otimes I_s(g))$$

est celui obtenu en considérant $I_s(g)$ comme *sous-algèbre* de $I_\Lambda(G) \otimes I_s(g)$, tandis que l'homomorphisme

$$H(I_\Lambda(G) \otimes I_s(g)) \longrightarrow I_\Lambda(G)$$

est celui obtenu en considérant $I_\Lambda(G)$ comme algèbre quotient de $I_\Lambda(G) \otimes I_s(g)$ par l'idéal engendré par $I_s^+(g)$.)

Corollaire. — L'algèbre de cohomologie $H(G/g)$ et les homomorphismes $I_s(g) \longrightarrow H(G/g) \longrightarrow H(G)$ sont entièrement déterminés par la connaissance de l'homomorphisme $I_s(G) \longrightarrow I_s(g)$, qui caractérise ainsi la « position homologique » du sous-groupe g dans le groupe G. Signalons les deux cas extrêmes :

1° Celui où l'homomorphisme $I_s(G) \longrightarrow I_s(g)$ applique $I_s(G)$ *sur* $I_s(g)$: c'est le cas, classique, où g est non homologue à zéro dans G ;

2° Celui où l'homomorphisme $I_s(G) \longrightarrow I_s(g)$ est *biunivoque* : c'est le cas où les *rangs* $r(g)$ et $r(G)$ sont égaux.

Dans tous les cas, le théorème 5 montre que l'algèbre de cohomologie $H(G/g)$ est justiciable de la théorie de Koszul concernant l'« homologie des S-algèbres » (voir la conférence de Koszul) : $H(G/g)$ s'identifie à l'« algèbre d'homologie de la S-algèbre $I_s(g)$ », S désignant ici l'algèbre $I_s(G)$, et la structure de S-algèbre de $I_s(g)$ étant définie par l'homomorphisme de $I_s(G)$ dans $I_s(g)$.

10. Quelques résultats concernant la cohomologie des espaces homogènes

Signalons, sans démonstration, une série de résultats qui se déduisent assez facilement de ce qui précède. Nous écrirons, pour abréger, $I(G)$ et $I(g)$ au lieu de $I_s(G)$ et $I_s(g)$. On notera $I^+(G)$ l'idéal de $I(G)$ formé des éléments de degré > 0 ; définition analogue de $I^+(g)$. Enfin, on écrira $P(G)$ au lieu de $P_\Lambda(G)$.

Écrivons la suite des homomorphismes canoniques

$$I(G) \longrightarrow I(g) \longrightarrow H(G/g) \longrightarrow H(G) \longrightarrow H(g) .$$

Le noyau de l'homomorphisme $I(g) \longrightarrow H(G/g)$ *est l'idéal J engendré, dans l'algèbre $I(g)$, par l'image de $I^+(G)$. Donc la sous-algèbre caractéristique de $H(G/g)$ est canoniquement* isomorphe à l'algèbre quotient $I(g)/J$. Rappelons que ses éléments sont de *degrés pairs*.

L'*image* de $H(G/g)$ dans $H(G)$ est une sous-algèbre (que nous noterons $H_g(G)$) engendrée par un sous-espace $P_g(G)$ de l'espace $P(G)$ des éléments primitifs de $H(G)$ [1]. On obtient $P_g(G)$ de la façon suivante : la différentielle δ applique $P(G)$ sur un sous-espace V de $I(g)$; soit J' l'idéal de $I(g)$, formé des combinaisons linéaires d'éléments de V à coefficients dans $I^+(g)$; J' est contenu dans J et indépendant du choix de δ. Alors $P_g(G)$ *est le sous-espace de* $P(G)$ *formé des éléments que* δ *applique dans* J'.

La dimension de l'espace vectoriel $P_g(G)$ est *au plus égale à la différence* $r(G) - r(g)$ *des rangs de* G *et de* g. D'autre part, l'image de $I^+(g)$ dans $H(G/g)$ est toujours contenue dans le noyau de l'homomorphisme $H(G/g) \longrightarrow H(G)$; pour que *ce noyau soit exactement l'idéal engendré par les éléments de degré* > 0 *de la sous-algèbre caractéristique*, il faut et il suffit que

$$\dim P_g(G) = r(G) - r(g) \ . \tag{3}$$

La condition (3) est trivialement vérifiée si $r(g) = r(G)$; dans ce cas, $H(G/g)$ est canoniquement isomorphe à $I(g)/J$. Elle est aussi vérifiée quand l'espace homogène G/g est *symétrique* (au sens de E. Cartan). Chaque fois qu'elle est vérifiée, $H(G/g)$ s'identifie au produit tensoriel d'algèbres

$$(I(g)/J) \otimes H_g(G) \ ,$$

et on a, d'après Koszul, une « formule de Hirsch » qui donne le polynôme de Poincaré de $I(g)/J$ (cf. conférence de Koszul) connaissant les polynômes de Poincaré de $H(G)$ et de $H(g)$, ainsi que les degrés des éléments primitifs de $H_g(G)$, on trouve immédiatement le polynôme de Poincaré de $H(G/g)$.

Si $\dim P_g(G) < r(G) - r(g)$, l'algèbre $H(G/g)$ est encore isomorphe à un produit tensoriel $K \otimes H_g(G)$, mais la structure de l'algèbre K est plus compliquée que lorsque (3) a lieu : K contient alors, outre une sous-algèbre isomorphe à $I(g)/J$, des générateurs de *degré impair*. Signalons qu'il existe des cas simples (J. Leray, A. Borel) où $r(g) < r(G)$, et où néanmoins $H_g(G)$ est réduit à 0.

En application des résultats relatifs au cas où (3) a lieu, on peut déterminer explicitement les polynômes de Poincaré des grassmanniennes réelles $G_{n, N}$ (il s'agit des grassmanniennes « orientées » : $G_{n, N}$ désigne l'espace des sous-espaces vectoriels

---

[1] Théorème connu de SAMELSON, *Ann. of Math.*, 42, 1941, Satz V.

orientés de dimension n dans l'espace numérique de dimension $N + n$). Posons

$$Q(t, p) = \prod_{1 \leq k \leq p} (1 - t^{4k}).$$

Alors le polynôme de Poincaré de la grassmanienne $G_{n, N}$ est :

si n et N sont pairs :

$$\frac{1 - t^{n+N}}{(1 - t^n)(1 - t^N)} \frac{Q\left(t, \frac{n+N}{2} - 1\right)}{Q\left(t, \frac{n}{2} - 1\right) Q\left(t, \frac{N}{2} - 1\right)}$$

si n est pair et N impair :

$$\frac{1}{1 - t^n} \frac{Q\left(t, \frac{n+N-1}{2}\right)}{Q\left(t, \frac{n}{2} - 1\right) Q\left(t, \frac{N-1}{2}\right)}$$

si n et N sont impairs :

$$(1 + t^{n+N-1}) \frac{Q\left(t, \frac{n+N}{2} - 1\right)}{Q\left(t, \frac{n-1}{2}\right) Q\left(t, \frac{N-1}{2}\right)}.$$

Bibliography

[AB]   M. F. Atiyah and R. Bott, *The moment map and equivariant cohomology*, Topology **23** (1984), no. 1, 1–28.

[AJ]   M. F. Atiyah and L. Jeffrey, *Topological Lagrangians and cohomology*, J.Geom.Phys. **7** (1990), no. 1, 119–136.

[AK]   L. Auslander and B. Kostant, *Polarization and unitary representations of solvable Lie groups*, Invent.Math. **14** (1971), 255–354.

[AM]   M. F. Atiyah and I. G. Macdonald, *Introduction to commutative algebra*, Addison-Wesley, Reading, MA. (1969).

[AMM]   A. Alexeev, A. Malkin and E. Meinrenken, *Lie group valued moment maps* (to appear).

[AP]   C. Allday and V. Puppe *Cohomological Methods in Transformation Groups*, Cambridge Stud. in Adv. Math. **32** Cambridge University Press, Cambridge, (1993).

[At]   M.F. Atiyah, *Convexity and commuting Hamiltonians*, Bull. London Math. Soc. **14** (1982), no. 1, 1–15.

[Be]   F.A. Berezin, *Introduction to Superanalysis*, Math. Phys. and Appl. Math., **9** D. Reidel Publishing Co., Dordrecht-Boston (1987).

[BGV]   N. Berline, E. Getzler and M. Vergne, *Heat Kernels and Dirac Operators*, Fund. Princ. of Math. Sci. **298** Springer–Verlag, Berlin (1992).

[Bo]   A. Borel, *Seminar on transformation groups*, Annals of Math. Studies. **46**, Princeton University Press, 1960.

[Bo2]   A. Borel, *Sur la cohomologie des espaces fibrés principaux et des espace homogènes des groupes de Lie compacts*, Ann. of Math. **57** (1953) pp. 115–207

[Bott]   R. Bott, *Vector fields and characteristic numbers*, Michigan Math.J. **14** (1967), 231–244.

222   Bibliography

[Bour]   N. Bourbaki, *Éléments de Mathématique*, Livre III, Topologie Générale, Chaptre 1, Actualit. Sci. et Ind. **1142**, Hermann, Paris (1965).

[BT]   R. Bott and L. W. Tu, *Differential forms in algebraic topology*, Grad. Texts in Math., **82** Springer–Verlag, New York - Berlin, 1982.

[BV]   N. Berline and M. Vergne, *Classes caractéristiques équivariantes. Formules de localisation en cohomologie équivariante*, C.R. Acad. Sci. Paris Sér. I Math. **295** (1982), 539–541.

[C]   S.S. Chern *An elementary proof of the existence of isothermal parameters on a surface*, Proc. AMS **6** (1955), 771-782

[Ch]   S.J. Cheng, *Representations of central extension of differentiably simple Lie superalgebras*, Comm.Math.Phys. **154** (1993), no. 3, 555–568.

[Chev]   C. Chevalley, *Invariants of finite groups generated by reflections*, Amer.J. Math. **77** (1955), 778–782.

[CS]   S. Cappell and J.L.Shaneson, *Genera of algebraic varieties and counting lattice points*, Bull.Amer.Math.Soc. **30** (1994), no. 1, 62–69.

[CS2]   T. Chang and T. Skjelbred, *The topological Schur lemma and related results*, Ann. of Math. **100** (1974), 307–321.

[D]   H. Duistermaat, *Equivariant cohomology and stationary phase*, preprint **817**, Univ. of Utrecht (1993).

[Dan]   V. Danilov, *The geometry of toric varieties*, Russian Math.Surveys **33** (1978), 97–154.

[Del]   T. Delzant, *Hamiltoniens périodiques et image convex de l'application moment*, Bull.Soc.Math.France **116** (1988), no. 3, 315–339.

[DGMW]   H. Duistermaat, V. Guillemin, E. Meinrenken and S. Wu, *Symplectic reduction and Riemann-Roch for circle actions*, Math.Res.Lett. **2** (1995), no. 3, 259–266.

[Du]   M. Duflo, *Théorie de Mackey pour les groupes de Lie algébriques*, Acta. Math. **149** (1982), no.3-4, 153–213.

[DV]   M. Duflo and M. Vergne, *Cohomologie équivariante et descente*, Astérisque **215** (1993), 5-108.

[dW]   B. de Witt, *Supermanifolds*, Cambridge Univ. Press, Cambridge (1992).

[GHV]   W. Greub, S. Halperin, and R. Vanstone, *Connections, Curvature, and Cohomology, Vol. III*, Academic Press, New York, San Francisco, London (1976).

[Gi]   V. Ginzburg, *Equivariant cohomologies and Kaehler geometry*, Functional.Anal.Appl. **21** (1987), no.4, 271–283.

[GKM]   M. Goresky, R. Kottwitz and R. MacPherson, *Equivariant cohomology, Koszul duality and the localization theorem*, Invent.Math. **131** (1998), no. 1, 25–83.

[GLS]   V. Guillemin, E. Lerman and S. Sternberg, *Symplectic Fibrations and Multiplicity Diagrams*, Cambridge Univ. Press, Cambridge (1996).

[GLSW] M. Gotay, R. Lashof, J. Sniatycki and A. Weinstein, *Closed forms on symplectic fiber bundles*, Comment.Math.Helv. **58** (1983),no. 4, 617–621.

[Go]   R. Godement, *Topologie algébriques et théorie des faisceaux*, Actual. Sci. et Ind. **1252**, Hermann, Paris (1958).

[GPS]   V. Guillemin, E. Prato and R. Souza, *Consequences of quasi-free*, Ann.Global Anal.Geom. **8** (1990), no. 1, 77-85.

[GS]   V. Guillemin and S. Sternberg, *Symplectic Techniques in Physics*, Cambridge Univ. Press, Cambridge (1984).

[Gu1]   V. Guillemin, *Moment maps and combinatorial invariants of Hamiltonian T^n-spaces*, Progress in Mathematics **122**, Birkhäuser, Boston (1994).

[Gu2]   V.Guillemin, *Riemann-Roch for toric orbifolds*, J.Diff.Geom. **45** (1997), no.1, 53–73.

[GW]   V. Ginzburg and J. Weitsman, *Lie-Poisson structures on some Poisson Lie groups*, JAMS 5 (1992), 445–453.

[Helg] S. Helgason, *Differential Geometry and Symmetric Spaces*, Academic Press, New York (1962).

[Hs]   W. Y. Hsiang, *Cohomology theory of topological transformation groups*, Erg.Math **85**, Springer–Verlag, Berlin, (1975).

[Je]   L. Jeffrey, *Group cohomology construction of the cohomology of moduli spaces of flat connections on 2-manifolds*, Duke Math.J. **77** (1995), no. 2, 407–429.

[JK]   L. Jeffrey and F. Kirwan, *Localization for non-abelian group actions*, Topology **34** (1995), no. 2, 291–327.

[Ka]   J. Kalkman, *A BRST Model Applied to Symplectic Geometry*, Thesis, Utrecht (1993).

[Ki]   A. Kirillov, *Unitary representations of nilpotent Lie groups*, Uspekhi Mat.Nauk. **17** (1962), no. 4, 57–110.

[Ko1]   B. Kostant, *Graded manifolds, graded Lie theory and prequantization*, Lecture Notes in Math. **570**, Springer–Verlag, Berlin (1977), 177–306.

[Ko2]   B. Kostant, *Quantization and unitary representations*, Lecture Notes in Math. **170**, Springer–Verlag, Berlin (1970), 87–208.

[Kos]   J.L. Koszul, *Homologie et cohomologie des algèbres de Lie*, Bull. Soc. Math. France., **78** (1950), 65–127.

[Ku]   N. Kuiper, *The homotopy type of the unitary group of Hilbert space*, Topology **3** (1965), 19–30.

[KV]   S. Kumar and M. Vergne, *Equivariant cohomology with generalized coefficients*, Astérisque **215** (1993), 109–204.

[Li]   Z. Li, *The Mackey obstruction and the coadjoint orbits*, Trans. Amer. Math. Soc. **346** (1994), no.2, 693–705.

[Ma]   W. S. Massey, *Exact couples in algebraic topology* I,II: Ann. of Math. **56** (1952), 363-396; III,IV,V: Ann. of Math. **57** (1953), 248–286.

[Mat]   O. Mathieu, *Harmonic cohomology classes of symplectic manifolds*, Comment.Math.Helv. **70** (1995), no. 1, 1–9.

[Me1]   E. Meinrenken, *On Riemann-Roch formula for multiplicities*, J.Amer.Math.Soc. **9** (1996), no. 2, 373–389.

[Me2]   E. Meinrenken, *Symplectic surgery and the Spin^C Dirac operator*, Adv.Math. (to appear).

[Mi]   J. Milnor, *Construction of universal bundles, I and II*, Ann. of Math **63** (1956), 272–284 and 430–436.

[MQ]   V. Mathai and D. Quillen, *Superconnections, Thom classes and equivariant differential forms*, Topology **25** (1986), no. 1, 85–110.

[MS]   J. Milnor and J. Stasheff, *Characteristic Classes*, Ann. of Math. Stud., **76**, Princeton Univ. Press, Princeton, NJ (1974).

[Qu]   D. Quillen, *Superconnections and the Chern character*, Topology **24** (1985), no. 1, 89–95.

[So]   J.M. Souriau, *Quantification géométrique*, Comm.Math.Phys. **1** (1966), 374–398.

[Sp]   E.H. Spanier, *Algebraic Topology*, McGraw-Hill, New York (1966).

[St1]   S. Sternberg, *Minimal coupling and the symplectic mechanics of a classical particle in the presence of a Yang-Mills field*, Proc.Nat.Acad.Sci.USA **74** (1977), no. 12, 5253–5254.

[St2]   S. Sternberg, *Symplectic homogeneous spaces*, Trans. Amer. Math. Soc. **212** (1975), 113–130.

[SU]   S. Sternberg and T. Ungar, *Classical and prequantized mechanics without Lagrangians or Hamiltonians*, Hadronic J. **1** (1978), no. 1, 33–76.

[TW]   S. Tolman and J. Weitsman, *The cohomology rings of abelian symplectic quotients*, (to appear).

[VdW]   B.L. Van der Waerden, *Modern Algebra*, Frederick Ungar, New York (1949).

[Ve]   M. Vergne, *Quantification géométrique et multiplicités*, C.R. Acad. Sci. Paris Sér. I Math. **319** (1994), 327–332.

[W]   A. Weinstein, *A universal phase space for particles in a Yang-Mills field*, Lett.Math.Phys. **2** (1977/1978), no. 5, 417–420.

[We]   A. Weil, *Sur les théorèmes de DeRham*, Comment.Math.Helv. **26** (1952), 119–145.

[Wi]   E. Witten, *Two dimensional gauge theories revisited*, J.Geom.Phys. **9** (1992), no. 4, 303–368.

[Wo]   N. Woodhouse, *Geometric Quantization*, The Clarenden Press, Oxford University Press, New York(1992).

[Zi]   F. Ziegler, *Thèse*, Université de Marseille (1997).

Index

Springer
and the
environment

At Springer we firmly believe that an international science publisher has a special obligation to the environment, and our corporate policies consistently reflect this conviction.

We also expect our business partners – paper mills, printers, packaging manufacturers, etc. – to commit themselves to using materials and production processes that do not harm the environment. The paper in this book is made from low- or no-chlorine pulp and is acid free, in conformance with international standards for paper permanency.